# Beginner's Guide To Bigfoot Research

By
Joedy Cook

Copyright © 2011 By Joedy Cook

ISBN9781461118268
Printed in USA

# Chapters

**Chapter 1: What Is Bigfoot? – pg. 9**

**Chapter 2: Where Do I Look? – pg. 26**

**Chapter 3: Bigfoot Investigation – pg. 40**

**Chapter 4: Description Of A Bigfoot – pg. 63**

**Chapter 5: Track Casting – pg. 82**

**Chapter 6: Investigation Techniques – pg. 87**

**Chapter 7: Bigfoot Hunting Techniques – pg. 92**

**Chapter 8: Equipment – pg. 100**

**Chapter 9: Bigfoot websites and forums-pg. 141**

**Chapter 10: Bigfoot Names-pg. 145**

in loving memory of my daughter Samantha H. Cook
Born July 21, 1995 Died July 13, 2010

The saddest word that mankind knows will always be "goodbye"
and so when little ones depart we who are left behind must realize
how much god loves little children, for angels are hard to find

**Special Thanks to**

**Ken Gerhard**
**Austin Grow**
**Bigfoot 101**
**Pete Travers**
**Melissa Hovey**
**TristateBigfoot.com**
**Bigfoothunt.com**
**Jeff Morris**

**By: Joedy Cook**

# Introduction

The pursuit of Bigfoot is fascinating. There is something that is incredibly interesting and exciting about the entire field. On one side of the coin, your goal has become to scientifically discover something that most scientists swear doesn't exist. Despite all of the evidence that exists that supports the existence of this creature, the world as a whole doesn't believe in it. Parents will often times tell their children that there is "no such thing" as Bigfoot. Scientists will ostracize the fringe element that devotes their time to pursuing the creature. What I'm trying to say is that one aspect that makes this field exciting is the constant possibility that you will be able to prove all of these people wrong. There is a chance, every time you go out into the field that you will come across a body of the creature or find some other irrefutable proof that there are large primates other than humans on the North American continent.

Along the same lines, more excitement comes through the fact that if you were to find this irrefutable proof of the creature, you will be forever famous. When real scientists begin studying the creature or when parents begin telling their children that there "is such a thing" as Bigfoot, your name will always be integral to that fact if you are the one who finally discovers it. So, one aspect of this excitement comes from the possibility with every investigation that you can convince the majority of the world that they are wrong.

The other side of the coin is equally exciting but in a different way. Bigfoot does not want to be found. It keeps to the shadows and is notorious for hiding itself away from humanity most of the time. The other form of excitement that is apparent in this field is the battle that is constantly being waged by the Bigfoot researcher. It is the Bigfoot researcher against nature. Humanity and nature have never mixed very well. There are a million things out in the wilderness that can hurt or even kill you. There are poisonous snakes all over the continent. Large mammals like bears and mountain lions have attacked and killed people. There are even stories about Bigfoot creatures who attack humans. Beyond these dangers, there are always dangers such as getting lost or falling and getting hurt especially when you are in remote areas of wilderness.

While the constant danger when a researcher is pitted against nature may seem like a turn off, we believe that it can also add to the thrill and excitement of the search for Bigfoot. This book contains many rules that you should follow when you are out in the wilderness, rules that help to keep you safe, but there are always unexpected things that nature can throw at you. In this way, Bigfoot researchers need to be resourceful and brave in order to be successful in the wild. So not only is there a chance that you could find fame and fortune by discovering proof of the creature, you also have the constant thrill and excitement of defeating nature's inhospitality every time you go out into the field and come back unscathed.

At the same time that this field is incredibly exciting to those involved, this same excitement can create some very real pitfalls that you need to look out for. Perhaps the biggest pitfall that has the tendency to afflict those who become obsessed with their search for the creature, is the fact that some researchers may allow their obsession with this to overcome their devotion to their families.

We feel that this in unacceptable. Family should always come first. Although it is important to spend a considerable amount of time in the search for Bigfoot in order to become successful, it is an unacceptable sacrifice to make to sacrifice the solidarity and happiness of your family in that pursuit. Too many researchers have spent so many hours out in the field that they were unable to realize that they were neglecting their own families. With the excitement that comes through the hunt for the creature, it is sometimes difficult to prioritize your family, but we feel that family is the most important thing. We will always respect those researchers who cannot make it to an investigation because they feel that they need to spend that time with their families. Family is the best excuse you can give, and it is an excuse that you should give if you feel that you have been neglecting them.

Next, it is important that you put your job before your research into the existence of Bigfoot. Not only do you need money to eat and live, but you also need money in order to purchase equipment and actually mount your search for the creature. It is very difficult to make money being a Bigfoot researcher. Those few of us who are lucky enough to be able to make a living at it have been in the field for many years. People know our names so they invite us to speak at conventions. We have many books out on the subject. In short, we spent a very long time getting to where we are and we had real jobs and kept our real jobs as we acquired our notoriety in the field.

So because the Bigfoot phenomenon is such an exciting field of study, it comes with several pitfalls that are quite important to avoid. In our minds there are many things that are more important than a blind pursuit of the creature. First of all, your own safety is more important than the discovery of the creature. Secondly, family is more important. Thirdly, your own livelihood and career are more important.

This being said, I am excited to begin our *Beginner's Guide to Bigfoot Research.*

# Chapter One

# What is a Bigfoot?

In the search for the creature that has become known as Bigfoot across the continent, it is important to first discuss the particulars of what you will be looking for. We need to talk about what a Bigfoot looks like, how a Bigfoot might behave, and how Bigfoot may differ in appearance and behavior across the continent. After all, as a Bigfoot researcher, you need to know whether or not what you are seeing is actually a Bigfoot and not some other creature.

It is quite difficult to talk about exactly what a Bigfoot might look like and how it may behave. One reason for this is that no one has actually captured or found the body of the creature in the wild. All of the information in this chapter therefore has to be based upon eyewitness accounts from people who claim to have actually seen the creature. Unfortunately, these accounts vary drastically. The color of the creature's fur, the height, the aggressiveness, and many other things will all vary depending upon the witness and the part of the country in which the creature was spotted. In order to overcome these differences and try to create an accurate depiction of the creature, we will split this chapter into several sections.

In the first section of this chapter we will examine the similarities between the Bigfoot creatures that have been reported across the continent. In the next section of the chapter we will attempt to examine the differences and categorize these differences by geography. What we mean by this is that descriptions of the creature in the western part of the continent vary greatly from descriptions of the creature in the eastern part of the continent, so we will point out these differences in order to help you, the new researcher, realize what kind of creature you are looking for dependant upon what part of the continent you are looking in. Beyond this, there are certain smaller geographic regions which have completely different descriptions of the creature. For example, Ohio has the Grassman, and Florida has the skunk ape. Many other small geographic regions have their own names and descriptions of the creature and we will attempt to describe these creatures to the best of our abilities.

**Things that aren't a Bigfoot**

The first thing that you are looking for when determining whether a creature is a Bigfoot, is whether or not the creature is a primate. The fact that the creature is a primate does not automatically make the creature a Bigfoot though. There is one primate that is indigenous to the North American continent, and there are many other primates who live on this continent today.

      Human beings are primates and have been on this continent for more than ten thousand years. There have been human beings in the past that have been more than eight feet tall like the giant Robert Wadlow who lived in the early 1900s. Wadlow stood a full eight feet eleven inches tall so was of a comparable size to the Bigfoot that have been reported around the country. Wadlow was definitely human though and did not resemble a Bigfoot in any way beyond his stature though. Other humans have been known to grow hair all over their bodies. For example, a man named Bassou from the Valley of Dades in Morocco was completely covered in hair. For this reason, he was expelled from his tribe and called an ape man in 1937. Again though, he did not resemble the descriptions of Bigfoot beyond the fact that his body was completely covered by hair. Other humans such as Julia Pastrana from America and Zana from Russia were also completely covered by hair but were definitively human beings. So the fact that a living being is giant and is covered by hair doesn't necessarily preclude that being from being human. There are other facets that need to be examined. Beyond this, real life creatures such as gorillas are large hairy primates.

Some may argue that there are no gorillas or non-human primates on the North American continent. While it is true that there are no non-human primates which are indigenous to the North American continent, there are in fact many here today. Hundreds of zoos across the country have exhibits with gorillas and many other non-human primates on display. From time to time, creatures kept in zoos in captivity have been known to escape into the wild. Also, many states throughout the United States have no laws limiting what kind of exotic pets you may own. There are many people throughout the country who have purchased baby primates but soon realized that they were unable to care for the adult creatures once they grew up. Pet owners such as this will often times release these primates for whom they can no longer care into the wild. Since escapes and releases like these are quite rare and a rather large group would be necessary in order to breed and sustain a population in the wild, it is nearly impossible that a group of escaped zoo gorillas or pet gorillas are now living independently in the wild. This being said though, it is not impossible for someone to one day run across an escaped gorilla in the wild. In this case it is important that you know the differences between the supposed Bigfoot creatures and the scientifically verified African gorilla.

In the next couple sections of this book we will examine the similarities and differences between the creature and other known creatures around the world. We will also examine similarities and differences between Bigfoot creatures themselves.

**Traits Similar to All Bigfoot Creatures**

1) *They are covered with hair*
   There are many things about the hair which vary from creature to creature. For example, the length of the hair will be different depending upon what part of the continent in which you see the creature. In the same way the color of the hair varies wildly throughout the country.
   The only thing that is uniform throughout all the sightings of the creature on the continent is that they are all covered with hair. Every part of their body has some kind of hair on it. The arms, legs, back, chest, head, and sometimes the face are all covered with some sort of hair or fur. This is an

important distinction when differentiating the creature from a human being. Although a human who has been living in the wild (a phenomenon which has been documented throughout history, although, admittedly, it has become increasingly rare in modern times) could be immense in stature and may share many other traits with the creature, it is rare that humans are completely covered by thick hair. As mentioned in the previous section though, it is not impossible that a human be covered with hair. That is why it is important that you read on and become familiar with other facets that are shared by all Bigfoot creatures.

2) *They have a strong butt*
This may sound funny at first, but this trait is a very important difference between Bigfoot and every other non-human primate in the world. By having a 'strong butt' we mean strong muscles like those that make up a human's butt: the gluteus maximus muscles. The butt muscles in primates help them to hold their body upright. In primates such as baboons and gorillas, the gluteus maximus muscles help the creatures to sit upright unlike many other species of animals around the world. These muscles in non-human primates are weak though when compared to human butt muscles. The butt muscles in humans allow for them to walk upright. Other primates are what are known as knuckle-walkers. They cannot stand upright when they walk, so

they therefore need to use their knuckles to support the weight of their upper bodies.

Bigfoot creatures, according to the sightings, do not have this trait in common with other primates. They are never seen walking on their knuckles. Instead, they are always seen walking upright like a human would. This is often times supported by tracks that have been found of the Bigfoot creature. There are footprints that are found, but there are never marks where the creature may have used its knuckles in order to support its weight. Logically then, since the Bigfoot is always seen walking upright like a human, we can assume that Bigfoot has strong butt muscles with which to support its upper body.

3) *They are large mammals*
The actual size of these creatures varies greatly from sighting to sighting, especially from east coast to west coast. There is a span that the height and weight of these creatures generally seems to fall within though. The creature almost always seems to be more than five feet tall and is always less than nine feet tall. The weight of the creature is estimated to be generally between 200 and 700 pounds. We realize that this is a huge span of height and weight, but there is a significant number of creatures that these sizes precludes. This cannot be a small primate or monkey. In fact, there are very few primate

species which fit within these height and weight requirements.

While most reports of Bigfoot suggest that the creature is larger than an average human or gorilla, humans and gorillas will sometimes fit within the height and weight requirements of the creature. Also, bears will fit into these size requirements, and more importantly, bears standing on their hind legs will often times resemble a Bigfoot. When observing the creature (or more importantly when observing evidence left by the creature), be careful that you examine the possibility that the creature that you are seeing or tracking isn't simply a human being or a bear.

4) *They have human-like feet*
While it is incredibly rare that a witness to the creature may actually see the creature's foot, it is actually rather common that researchers and investigators find a Bigfoot footprint while they are searching for the elusive creature. In fact, the name Bigfoot itself came from the discovery of footprints by a construction crew in Bluff Creek, California. The work crew discovered a series of footprints that looked surprisingly human. The only difference that the work crew could tell between these prints and actual human prints were the sheer size of these prints. They made casts of the prints and brought them to the *Humbolt Times* newspaper in Eureka, California. When the story broke, they referred to the

creature that made the prints as Bigfoot since it resembled a normal human foot, except that it was, well, big.

So when comparing Bigfoot feet to human feet there are a couple things to keep in mind. First of all, Bigfoot feet are generally much larger than any human foot out there. The toes, like those of a human, are all on the front of the foot. There is a pronounced heel and a ball of the foot in any Bigfoot casts that have been made. The biggest difference between human prints and Bigfoot prints, other than the size, is that Bigfoot prints tend to be flatter. What I mean by this is that when comparing the print of a human to that of a Bigfoot, the human print will curve inward dramatically between the ball of the foot and the heel on the inside of the print. Bigfoot tracks will often not do this at all. Bigfoot tracks will often go straight back from the ball of the foot to the heel. Despite the difference of size and the slight difference in shape, you can still mistake a Bigfoot print for a human print. Another thing to keep in mind when examining a track is the location in which you found it. Humans are unlikely to walk barefooted through deep wilderness or snow so if you have a print that you are unsure of, consider the possibility that a human would be barefooted in that area. On that same note, there is always the possibility that the track was made by a hoaxer, an unfortunate phenomenon that we will discuss in depth in a later chapter.

There are other animals that made prints that are similar to Bigfoot feet. These prints have pronounced differences though that should be discussed here. The first type of tracks we would like to discuss are large, non-human primate tracks that aren't Bigfoot's. Gorillas, chimpanzees, and orangutans are all large primates whose prints could be mistaken for that of a Bigfoot. (The chances of a gorilla, chimpanzee, or orangutan being in the wild was discussed earlier in this chapter.) The way to tell the difference between a known large primate print and the print of a Bigfoot is in the configuration of the toes. As previously mentioned, in a Bigfoot print the toes are all aligned at the front of the foot. On the other hand, in primates such as gorillas, chimpanzees, and orangutans the big toe is on the side of the foot like a thumb. This actually gives these known primate tracks the appearance of being handprints. So the orientation of the big toe is the easiest way to differentiate between a Bigfoot print and the print of another known primate.

Finally, it is possible to mistake the prints of a bear for those of a Bigfoot. Bear prints have five toes all situated at the front of the foot like those of a Bigfoot. This being true though, bear tracks are quite a bit shorter than Bigfoot tracks. It is nearly impossible to mistake a single bear track for that of a Bigfoot due solely to length. Unfortunately there is a phenomenon that could exist that may could lead to a bear track being mistaken for a Bigfoot track. This phenomenon is called double-tracking. What happens is the front paw of a bear creates a track and then the back paw steps partially into the first track. This creates a longer print that can appear somewhat like a Bigfoot track. There are a couple things to keep in mind while examining a print for the possibility of its being a double bear track. First, the front of the bear's foot is slightly wider than the back. Therefore, if you have a double track, most of the time the toes from the second track will spill over the edges of the heel of the first track. What we mean by this is that the track will have a somewhat hourglass shape, and you will often be able to see the toes of the creature sticking out from the middle of the track. Further, the toes of a bear are slightly different from those of a Bigfoot. Despite being on the front of the foot like those of a Bigfoot, the big toe is on the outside of a bear's foot and the toes follow more of an arch shape around the front of the foot instead of a slant shape like those of a human or Bigfoot.

**Traits that differ in Bigfoot from region to region**

Now that we have discussed the traits that all Bigfoot have in common, it is now proper that we discuss how Bigfoot differ from region to region. Depending upon what part of the country you are searching in for the elusive creature, it displays different traits. In the following few paragraphs, we will attempt to generalize the differences between the creatures throughout the country. This differences are only generalizations. It is possible that a creature from the east will display the traits of a creature from the west, but overall these generalizations tend to be true based upon the region in which the creature is sighted.

### 1) Size

Bigfoot tend to differ in size dependant upon the region in which they are sighted. When talking about differences in size, we will divide the Bigfoot creature into two regions. One size of creature lives west of the Rocky Mountains and another lives on the east side of the range.

The creatures that have been sighted on the western side of the Rocky Mountains tend to be larger than those on the eastern side. This includes creatures that have been spotted in states such as California, Oregon, and Washington and includes the majority of creatures that have been spotted. (We will discuss frequency of sightings and areas most likely to spot the creature in a later chapter.) Creatures in these states tend to be anywhere from about 7 feet tall to about 10 feet tall. While there have been human beings that have been 7 or 8 feet tall, there has never been a human that has been 9 feet tall or certainly not 10 feet tall.

On the eastern side of the Rockies, the Bigfoot that have been sighted are still large creatures. They range anywhere from about six feet tall to about seven or seven and a half feet tall. The further south that you move in the country the shorter the creatures tend to get. Creatures in Ohio tend to be closer to the seven to seven and a half feet tall mark while creatures in Florida, Arkansas, or Louisiana tend to be closer to the six foot mark.

This is not to say that all creatures in these regions necessarily fall into these height requirements. There have been creatures spotted in the west which have been estimated to be shorter than seven feet tall, and there have been creatures spotted in the east which have been taller than the seven and a half foot cap. What we mean to say here is that in general, the further northwest you move in the country, the larger the creatures tend to get.

*2) Demeanor*
Again when talking about the demeanor of the creature, we can separate them into those creatures west of the Rockies and those east of the Rockies. Again, like with size, the demeanor of these creatures seems to shift dramatically as you move across the country.

The Bigfoot that live in the Pacific Northwest, on the western side of the Rocky Mountains, tend to be more gentle and more reclusive. There has hardly ever been a report of a creature attacking a human being in the west. Usually, when the creature is sighted in the west, it quickly moves away and disappears deep into the wilderness. Also, in the west, the creature's seem to be quite a bit more shy. The sightings are almost never near urban centers and tend to be in deep wilderness areas. When the creatures are seen they tend to move towards the safety of the wilderness and away from the sight of humans. This doesn't mean that there have been less sightings of the creature in the west. In fact, the opposite is true. This is due to the fact that there is more untamed wilderness out in the west and likely due to this there is a much larger population of Bigfoot creatures. So while in the west the creature makes more of an effort to stay away from humans, the larger population of creatures makes for a higher volume of sightings.

    In the east, the demeanor of the creature is dramatically different. East of the Rockies, the creature is much more likely to interact and sometimes attack human beings. Violent encounters range from full fledged attacks where the creature actually starts hitting and assaulting humans to incidents where the creature throws rocks or sticks at hapless passersby. The creatures in the east are also more likely to live near urban centers. In Florida, the skunk ape mostly inhabits the Everglades. Some sightings of this creature are on the outskirts of Miami itself. Also throughout Ohio and Louisiana the creature has been seen in suburban areas just outside of large cities. While the creatures are definitely more aggressive than their western cousins, the reported sightings all seem to show that the creature is more curious than hateful towards mankind. Regardless of this though, it is probably in your best interest as an investigator to give the eastern creatures more space if you were to encounter one.

As a final point on the creature's demeanor, it seems that when talking about Bigfoot in the east, the creature seems to become more aggressive the further south you go. While in Ohio there have been many cases where the creature has attacked humans, it is far less common than it is in places like Arkansas, Louisiana, and Florida. Like it was with the size of the creature, the creatures in the northwest seem to be one extreme of the spectrum, the creatures in the southeast seem to be on the other side of the spectrum, and those creatures in Ohio, Pennsylvania, and Indiana seem to be almost a hybrid of these two extremes.

*3) Hair Color*
Like humans, the hair color of the creatures varies greatly, even within individual regions throughout the country. On the other hand though, there are basic trends in hair color that dictate that a creature from west of the Rocky Mountains is more likely to have a certain color of hair than a creature from the east.

In the west, the creatures tend to have a hair color like that of a bear. The color of the hair is almost always brown. Some of the creatures have a medium brown color, but most of the western Bigfoot creatures have a darker brown, almost black hair color. Perhaps the best piece of evidence supporting the existence of the creature, the Patterson film in which a Bigfoot creature is actually seen walking near a tree line, clearly shows the most common color reported in these western creatures. This dark brown color has become the generally accepted norm when talking about Bigfoot since so many of the creatures have been seen in the west, and so many of them have this dark brown hue.

East of the Rockies though, the creature will often times have a much different appearance in regards to color. In the Midwest and the Northeast, the creature is most often reported as having a reddish coloring. The creature will often times appear to have a rusty brown or red coat of fur. Further south in places such as Florida and Arkansas, the creature will have a grey coat of hair. The color of their hair will appear much like that of a raccoon or a squirrel. Other times the creature's hair will be a lighter grey, almost white color. This is almost exclusively a trait of the Bigfoot from the southeast.

Again, this is not to say that the creature will always have a particular color hair depending upon the region in which it is seen. These are just generalizations based upon the large volume of sightings of the creature across the country. This is just to say that more creatures in the Pacific Northwest have a dark brown coat of fur than any other color—just as the coat is more often rusty red in the Midwest and Northeast and grey in the south.

*4) Hair Length*

When talking about hair length, we need to divide Bigfoot into two categories into which we haven't yet divided them. We need to talk about Bigfoot from the north of the country versus Bigfoot from the south of the country.

In general, Bigfoot from the north of the country tend to have shorter hair. The hair is evenly distributed over most of their body and is very short, almost like a bear's fur. The only place on the body where this is not generally true in northern Bigfoot is on the forearms. Often times these northern Bigfoot will have long hair on its forearms very much like that of a gorilla.

In the south, on the other hand, the hair is usually long and shaggy all over the body. In Florida, Arkansas, Louisiana, and Texas especially, most sightings of the creature involve very long hair which hangs down all over the creature's body. The hair on the forearms isn't longer than the hair on the rest of its body like those in the north, but the hair on the southern Bigfoot's forearms is the same length generally as those in the north. Witnesses just do not notice a difference since all of the hair on the creature's body is of equal length.

These traits are the things that it is important for you to look for when you are investigating Bigfoot. When you go out in the field you cannot expect that you are going to see a Bigfoot. It is rare that investigators spot Bigfoot when they are researching out in the field. At the same time though, you can use the information that you have gained from this chapter in your investigations. For example, you can determine whether the footprint that you find is that of a Bigfoot or that of another known creature. Also, when you are interviewing witnesses, you should now have a better understanding of what a

Bigfoot is and whether or not the creature that these witnesses saw was actually a Bigfoot or if it was another creature that the witness simply misidentified.

This is not to say that you will never see a Bigfoot creature. There are many researchers out there who have actually seen the creature in the wild. In fact, perhaps the most famous sighting and best evidence for the existence of the creature, the Patterson sighting and resulting film, was made by a Bigfoot researcher. Patterson himself was a Bigfoot researcher who went out into the woods in northern California in order to find a Bigfoot. So while it is rare that people who are investigating sightings actually come across a Bigfoot, it is still possible that a researcher could actually come across the beast. Because of this, it is also important that you know the traits and behaviors to look for in the creature.

# Chapter Two

# Where Do I Look?

Since we now know what the creature we're looking for looks like, the next logical step in your training on how to become a Bigfoot researcher is finding out the best places to look for the creature. The best researchers in the field know that it is possible to find the creature pretty much anywhere in the country. At the same time though, they know that there are many places where the likelihood of coming across the creature is much higher.

In this chapter, we will attempt to discuss the best places for someone to search for the creature. We will examine this both in a general way and then in a more specific way, mentioning actual parks and tracts of wilderness where the creature is sighted most often. We will also look into specific sightings, explaining where the sightings took place and what exactly transpired during these famous Bigfoot sightings throughout the country.

*In what American states should I look for the creature?*

The answer to this question is actually quite a bit easier than it might seem. Unless you live in Hawaii, there has been a Bigfoot sighting in your state. There have been frequent sightings all across the country. Of course, there are those states in which these sightings are more frequent, but every state in the continental United States, Alaska, and most provinces in Canada have all had at least one sighting of the creature. Most of these states have had many sightings of the creature.

  If we were to discuss general trends in the appearance of the creature throughout the country, there are several states in which the likelihood of finding the creature is lower than in others. If you live in Hawaii, it is very unlikely that you will encounter the creature. There has never been a Bigfoot sighting in Hawaii for one. Also, it would be almost impossible for the creature to have made its way out to the island. Hawaii was never a part of the larger continental United States. It was formed in the middle of the ocean on a volcanic hot spot. Since the general consensus about where the creature came from is that it migrated to the North American continent over the Bering Land Bridge many thousands of years ago, there is no way that it could have ended up in Hawaii this way. Beyond this, since Bigfoot cannot swim hundreds of miles across the open ocean and they cannot construct and use boats, there seems to be no possible way that the creature could have traveled such a great distance over the ocean. So if you live in Hawaii, you will need to travel to the continental United States if you want any hope of glimpsing the creature.

  Even on the continental United States though, there are some areas where the creature is rarely seen. If you live in these states, it is possible that you may find some evidence or even catch a glimpse of the creature, but it is more unlikely than most anywhere else in the country. First I'll mention Delaware as being one of these states where sightings of the creature are incredibly rare. I mention Delaware by itself because it is kind of an anomaly. The other states where the creature seems to be rare are all grouped together in regions. Delaware stands alone. Delaware is surrounded by New Jersey, Maryland, and Virginia. These three surrounding states are relative hotspots for the creature. Remarkably though, we have only come across one single sighting of a Bigfoot creature in the entire state of Delaware. We don't mean to be misleading when we say the 'entire' state since Delaware is actually quite small. In fact, its size could very well be the reason that there are so few sightings in the state. Another reason that there are so few sightings in the state could be its location and the percentage of the state that is shoreline. Delaware runs rather vertically up the Atlantic coast and is not very wide. Because of this, there is not much land that is very far inland in the state. While the population of Delaware is quite small, most forested wilderness is further inland than the westernmost reaches of the state. Wilderness areas do exist within the state, and there has been at least one sighting of the creature in the state, but the creature is rare along the Atlantic shore in the northern half of the country. If you were in Delaware looking for the creature, I would suggest moving just outside the state's borders and searching in one of the surrounding states. You would be more likely to find evidence of the creature.

The next place that seems to have a marked lack of Bigfoot creatures is most of New England. While New York is one of the best states in the country to search for the creature and Maine also has a large number of Bigfoot sightings, the other states in New

England have very few sightings. This includes Vermont, New Hampshire, Massachusetts, Rhode Island, and Connecticut. This is rather strange because these states all have sparse populations and they all have rather large tracts of forested wilderness. Sparse population and forested wilderness are both almost prerequisites for a large population of Bigfoot creatures. So why is this area so sparsely populated by Bigfoot, and why are the surrounding states and provinces all heavily populated with the creature? It's hard to say. Perhaps since Europeans have been hunting and trapping in that area for such a long time, the Bigfoot have moved out of the area. This is to say that since the first settlers to the continent from Europe settled in this area, perhaps the Bigfoot that once inhabited the area have simply moved further west. Another theory is that perhaps the Bigfoot creatures just never settled in the area. The winters in the area are quite brutal and they are separated from the west by lakes, rivers, and mountains, so perhaps when the creatures originally migrated from the west, they simply did not settle in New England because of the harsh environment and natural barriers. At the same time though there have been some sightings of the creature in all of these states. Since the population is so sparse in these states, it is also possible that there is a large population of Bigfoot there, but there just aren't enough people to witness the creature. Due to the remote forested wilderness of the area, it is not a terrible place to launch a search for the creature, but again, if you were hoping to see a creature, the history of sightings there suggests that you will not have much luck.

Lastly, the area of the country that has very few sightings of Bigfoot that we haven't yet mentioned is the Great Plains states just east of the Rocky Mountains. Kansas has relatively few sightings, but the states that we are mostly talking about are North and South Dakota and Nebraska. All together, these three states have less than twenty reported sightings of the creature. This lack of sightings is somewhat easier to explain than the previous two regions that we have mentioned. These three states have very little of the forested wilderness which seems necessary to sustain these creatures. The states are filled with vast grassy or cultivated fields or with climates that don't promote heavy foliage. These states have an overall environment which is unlike that of many of the places where the creature seems to flourish.

Now that we have discussed the regions where you are least likely to find a Bigfoot creature, let's discuss the areas where you are most likely to discover the creature. To do this we will count through the top five states in the United States for Bigfoot sightings.

The state with the most sightings is California. This is where the Patterson film was shot and where Bigfoot tracks were first reported to the newspapers. Many sightings have occurred and still occur in the heavily forested northern parts of California. This is likely due to the vast stretches of wilderness which remain virtually untouched in the area. These vast tracts of secluded wilderness give the creature a place to live in secrecy, generally hidden from humanity.

The same is true for the second and third states on our list of the top five states for spotting this elusive creature. These next two states are also in the Pacific Northwest. They are Washington then Oregon. These areas also have vast tracts of wilderness that humans rarely wonder into. In fact it has been suggested that there is perhaps a higher population of these creatures in Washington and Oregon than in California. The reason that there have not been as many sightings is simply due to the fact that there are not as many people in Washington and Oregon as there are in California. The human population paradox states that there are more sightings in areas with higher human populations but that there are less creatures in places with higher human populations. What this means is that although there are less creatures in these places with higher human populations, there are more humans around who are there to report their sightings. So although there are less creatures, there are more people there to see these fewer creatures. So if you had the opportunity to chose any place in the country in which to search for the creature, any of these three states in the Pacific Northwest would probably be your best bet. Most of the most convincing evidence towards the existence of the creature has been collected from this area.

    The other two states on the top five list are states that people don't usually mention in relation to Bigfoot. Number four on the list is Ohio. While Ohio is one of the most populated states in the country, it also has vast areas that are still untamed wilderness. These areas are in the east and southeast of the state. The Bigfoot from Ohio has even acquired its own nickname, the Grassman. Hundreds of sightings have been reported throughout the state: usually to the northeast of Cincinnati, the east of Columbus, and south of Akron.

The final state in the top five is the state of Florida. This creature has also gained its own nickname and is called the Skunk Ape by locals because of its reportedly awful smell. The creature will often be sighted in the southern swampy areas of the state and has garnered hundreds of sightings throughout history.

All of the other areas of the country have fairly frequent sightings of the creature. The creature is seen rather often throughout the Rocky Mountains. It is seen in the swampy areas of Arkansas and Louisiana as well as the vast expanses of Texas. It is seen throughout the south and the Midwest as well as New York state, especially the Whitehall region in the northeast of the state.

So if you want to choose a state in which to search for the creature, chances are that the creature has been seen somewhere within the state in which you live (unless you live in Hawaii). So it is perfectly acceptable to mount your first investigation into the creature from the familiarity of your own state and climate. If you desperately want to see the creature and wish to maximize your chances of doing so, you should probably try to visit one of the five states that we have mentioned in this section or visit one of the top ten locations to search for the creature that will be mentioned in a later section of this chapter.

*Do Bigfoot like to live near people?*

This question is not as easy to answer as it might originally seem. The basic consensus among Bigfoot researchers is that Bigfoot are generally reclusive and that they try to stay as far away from humanity as possible. After all, all indications are that Bigfoot are able to sustain themselves without any aid from humanity. There have been many sightings of Bigfoot and rarely, if ever, do Bigfoot exhibit the traits of a scavenger. Bigfoot do not generally go through human dumpsters and garbage looking for discarded food. Only rarely do Bigfoot ever rely on humans for any kind of sustenance. The only examples of this are that sometimes Bigfoot are reported to attack caged livestock on a farm.

So why is this question so hard to answer then? We believe that the reason this question is so hard to answer is based on the human population paradox which we have introduced in the last section of this chapter. Most of the sightings of the creature are near the edges of human populations centers instead of in the deep untamed wilderness where these creatures are thought to hide.

Again, we feel that the reason for this occurrence is obvious. Sure, the creature mostly lives in the uninhabited deep wilderness areas of the United States. The reason that most of the sightings are at the edges of populated areas is simply because there are more people in these populated areas to see the creature. So while there are less creatures there, there are more people to see them so therefore there are more reported sightings near populations of people than in deep wilderness because there simply aren't people in the deep wilderness to see the creatures.

So what does this mean to you the Bigfoot researcher? Should you search in the outskirts of human populations where most of the sightings are, or should you search in the deep wilderness areas where researchers postulate that higher populations of the creatures live. Again, this is not an easy question. Most of the time, your best chance of getting evidence of the creature is to visit the site of a recent sighting. This way you know a very specific area in which to search. This will make your chances of finding evidence of the creature higher. At the same time if you travel into deep untraversed wilderness areas there are more likely to be higher populations of the creature. At the same time though, the areas are so vast and the populations still likely so small that your chances of encountering one will drop dramatically. You could improve these chances by finding game trails and places with fresh water, but if at all possible, visiting the site of a recent sighting would probably be your best chance at finding evidence of the creature.

*What environments can support the creature?*

There are many different environments across the country which seem to support Bigfoot creatures. The variety of environments seems to be due to the fact that the different races of Bigfoot have evolved in different climates and ecologic regions. For example, in general it seems like the northern Bigfoot tend to live in deep forested areas while the southern Bigfoot tend to inhabit wetlands or swampy areas. The Bigfoot in the north seem to prefer a colder climate while those in the south prefer a warmer climate. These differences in climate and ecologic regions though are based directly on the area in which the Bigfoot lives. In south Florida for example, the creature cannot live in a cold dense forest since most of south Florida is swampland. Therefore, these differences do not matter when you are looking for an environment which may support a Bigfoot creature. There are several things which seem to be common environmentally for most Bigfoot sightings.

    First of all, there needs to be large areas of untamed wilderness nearby. The creatures have never been captured and are rarely seen or photographed by humans, so they need a large area in which to hide. Small wooded areas that are completely surrounded by urban development should not be able to support a Bigfoot creature. Even if the Bigfoot has enough food and water in that small wooded area for a season or two, it will eventually have to migrate to a larger wilderness area when its resources have become extinguished. If the small wooded area is completely surrounded by development, it is unlikely that this creature could move across a suburban landscape without being seen and perhaps captured on film. It happens from time to time that one or two of the creatures will be in a small pocket of wilderness such as this. For example, just outside of Akron, Ohio, an impressive Bigfoot nest was discovered in a small pocket of woods that was completely surrounded by suburbs. The problem with searching small wooded areas such as this is that if these areas were inhabited frequently by the creatures, they would be seen more often as they migrated from these areas and through suburban neighborhoods.

So, on the one hand, Bigfoot need large tracts of wilderness in which they can live in anonymity. Since there has never been indisputable evidence supporting their existence, they would have to exist for the most part in very remote areas of the country. Besides this fact that the creatures most likely live in deep wilderness areas in order to hide from humanity, the environment in which they live also needs to have the resources available with which to support the creature.

Since the creature is reported to be quite large, it would therefore need a lot of food and water. When finding an environment that is suitable to support the creature, you should first confirm that it is connected to a large uninhabited wilderness and then should look for suitable freshwater sources. A river or freshwater lake would be necessary to the survival of such a creature. Beyond this, these bodies of water would be great places to mount your search for the creature. Several times every day, these creatures are likely required to drink water. Therefore, your best chance of catching evidence of the creature in the wild is likely to be near these sources of water.

Beyond water though, these creatures are in need of food. From evidence collected and sighting reports, it seems that these creatures are omnivores. Therefore, they subsist on both vegetation and meat. The creatures will likely eat roots and fruits throughout their environment and will also subsist on many small mammals like squirrels and raccoons. Some reports suggest that Bigfoot will also hunt deer from time to time, especially in the Midwest region of the country. So when you are looking for a suitable environment for Bigfoot, not only should you look for its remoteness and for water sources, but you should also look at the wildlife that is indigenous to the area. After all, Bigfoot need to eat.

*Where in the forest should I be looking?*

The first place that a researcher should always look for Bigfoot is near the site of a recent sighting. Whether or not this area seems to be a reasonable place to search environmentally, the area of a recent sighting is the most likely place to get results. The reason for this is obvious. Since the creature has been seen in that particular area, you now know that not only can the creature survive in that area but also that the creature has been in that area and so therefore may have left some evidence such as tracks or hair. Unfortunately, it is not always possible to follow up on recent sighting reports. If you are a beginner in the field, you are not often the first person to receive the reports of the sighting. If you receive these reports after other researchers, the other researchers may have already collected all of the evidence from the site. So while it is always best to follow up on any recent sightings of the creature, it is not always possible to do so.
   If you do not have recent witness reports and are going into a suitable environment essentially blind, there are several places throughout these forested regions

which are more likely to yield a sighting of the creature than others.

As we mentioned in the last section, the place that will perhaps be your best bet when searching for the creature is the local water source. Bigfoot need freshwater daily in order to survive, so if you feel that there are Bigfoot in a specific area, they will need to at one point or another visit these water sources.

Also, you may want to search for the creature along game trails. Many animals, which the Bigfoot will likely hunt, will traverse these trails throughout the wilderness. Where two or more of these trails intersect is also a very good place to search for the creature. This is because the Bigfoot creatures could possibly be using the same strategy when searching for their own prey. Since so many animals need to use these trails to travel through the thick woods, these trails are great places in which to mount a search for any creature out in the woods.

Since these creatures seem to be so elusive, it may seem logical that you move to the most lonely and remote corner of the wilderness to look for them. Like we mentioned here though, that is not necessarily true. These creatures are good at avoiding humanity, at the same time though, these creatures need to move alongside the majority of the wildlife out in this wilderness. So while it would be prudent to mount your search far away from human civilization, you should search alongside the wildlife in the area for the creature itself. Like the rest of the wildlife, the creature needs to drink, and the creature also needs to feed off of other creatures throughout the forest.

*What are the ten absolute best places to look for Bigfoot?*

1) Bluff Creek, California
2) Whitehall, New York
3) Mt. St Helens National Park, Washington
4) Everglades National Park, Florida
5) Salt Fork State Park, Ohio
6) Blue Rock State Forest, Ohio
7) Boggy Creek, Arkansas
8) Fondulac area, Minnesota
9) Raccoon Creek State Park, western Pennsylvania
10) Virginia

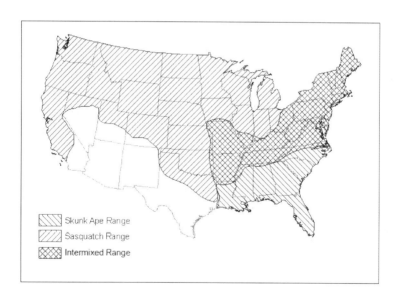

# Chapter Three

# Bigfoot Investigation: Step-By-Step

Each investigator has his or her own way of doing investigations. The way that we are telling you is not necessarily right or wrong; it is simply the way we feel is best. Feel free to use your own judgment when doing a Bigfoot investigation. We have put this step-by-step reference forward in order to share our own investigation techniques and to give a new researcher a guideline to follow during an investigation. Another way to learn to be the best investigator that you can be is to speak with other investigators in order to find out how they do their research and learn from that. Everyone has their own investigation techniques and even you, as a new researcher, will develop your own way of doing things.

      Before we get started with our description of how to mount a professional Bigfoot investigation, we feel that it is worthwhile to point something out. You are not necessarily going to come across some evidence of Bigfoot every time you go out into the field. Even when investigating those locations where a witness was involved, it is not a sure thing that you will come across some evidence. The best way to approach an investigation is to do so without expectations. You cannot go into an investigation thinking that you're going to find something because then you when you do find something, your mind will not be able to differentiate between genuine Bigfoot evidence and evidence of some other activity in the area. At the same time though, you cannot go into an investigation assuming that you will not find any evidence. If this were the case, you may discount evidence that may actually support the existence of the creature.

Approach a Bigfoot investigation like a CSI would approach a crime scene. Make no assumptions about any evidence that you are able to see and gather. Simply gather the evidence and make your own best scientific theories about what that evidence might be trying to say.

**The Bigfoot investigation**

Essentially you goal in a Bigfoot investigation is to collect all the evidence that you can as you are out in the field. We will go into more depth in a later chapter about exactly how you should collect the evidence, but here we want to be as detailed as possible on how to mount an actual investigation into the field.

There are many different things that may kick off an investigation. Many times, investigations are started when a witness claims to have seen the creature or seen some evidence in support of the creature. Since this is the most common type of investigation, we devoted a section into mounting that type of investigation. At the same time though, we realize that many beginners in the field may not have access to witnesses when they first get started in the field. Many times, researchers need to set up websites and phone lines to get these witnesses to report their findings. Beginners will often times not have the resources or the experience to have a successful website such as this and so therefore will not be able to consistently get first hand accounts from witnesses.

Since this is the case and this is a book for beginners, we have also included a section here which explains how to mount an investigation into any potential Bigfoot habitat. Before you start any investigation, there is something that you should keep in mind. Make sure that you always have all of your equipment with you at all times. If at all possible, keep your equipment in your car at all times so that it will be ready for you whenever you leave for an investigation. If that's not possible, at least have it packed up and ready to be picked up at a moment's notice. We'll talk much more about equipment in the next chapter.

**What if we get a witness who actually sees a Bigfoot?**

*Step 1: Check out the location*

If the location is close to where you live, actually go and check out the location as your first order of business. Get a feel for what the site looks like and look for any landmarks. It is important that you get a feel for what the area looks like because it would make it much easier for you to follow the story as the witness is telling the story to you. As the witness is trying to explain where it ran, it is much easier to understand if you have already seen the area that they're talking about.

    If the location is not near where you live, it may be better for you to look at the area by map instead of actually going all the way out to the location in the unfamiliar place. Again, if you look at a map and have a sense of the landmarks such as hills and creeks, it will be easier to understand the story that the witness is telling you.

    This is just a preliminary glance at the location. At this point you don't want to look for any footprints or any other evidence of the creature. You simply want to get a sense of your surroundings so that you can better understand your witnesses. You will have plenty of time to later look for evidence, but before you go running out into the field looking for evidence, you want to talk to your witnesses to get all of the information that you can before jumping out into the field.

*Step 2: Interview the witness(es)*

The next step in this particular type of investigation is interviewing your witnesses. This is important because you will likely get a lot of information that will be helpful further along in your investigation. They can tell you exactly where the saw the creature so that you know where to look for prints. They can tell you how tall the creature was so that you can look for broken branches and hair samples at the proper height. Any information about the sighting that you can garner from them could potentially be helpful to your investigation. Make sure you document all the facts that they tell you, no matter how trivial these facts might seem.

## Tips for conducting a successful interview

1) *Don't jump immediately into the sighting*
   When you are talking to the witness, it is very important that you work to build some sort of relationship with the witness. You want to first work to build a relationship with your interviewee. Make jokes, tell them that you like their clothes or their house, or participate in any kind of small talk that you can think of. If you don't jump right into the interview they will become more comfortable with you and will therefore open up more to you.

2) *Don't lead them into any questions*
   Another important thing to keep in mind is to not give the witness any answers. You want to hear the raw version of what they saw. You can of course ask them any questions that you want about the sighting, but make sure that you don't give them the answer in the question itself. For example, don't ask if the creature was brown, ask what color the creature was. Don't ask if it was down near the creek, ask where exactly on the map they saw the creature. If you give them a possible answer and they are foggy about any of the details, they will latch onto the detail that you let slip in the question and give you that one.

3) *Have them draw the creature*
   There is a lot of wisdom in the old adage which states that a picture is worth a thousand words. Even if the witness is a terrible artist, their bad drawing can often times be more descriptive of the creature that they saw than all the words they could possible use to describe it.

4) *Get all the details you can*
   As I mentioned before, every little detail can help. While at the time of the interview a little detail might feel worthless, you may find that out in the field it is invaluable. For example if the witness claimed that the creature seemed to trip slightly as it was walking, you may think that this was an unimportant detail in the story. It could prove valuable though as you are out in the field because you would know to look for prints or evidence near whatever it was that tripped up the creature in the general area of the sighting.

5) *Draw a small map of the area*
   While the witness is telling you the story of his or her sighting, draw a map of where you believe all of the events in the sighting took place. Since you have already scouted out the area briefly, you have an idea as to what the area looks like, so draw the map from your memory of the area. This could be important when you're out in the field because the map is drawn as you are listening to the story, so not only is the map drawn as you are hearing the report of the

sighting but the witness can also see the map as you're drawing it and correct it when needed.

6) *Record the interview*
You don't need to necessarily make an audio recording of the interview, but you do need to make sure that you are at least writing down everything that the witness says. Even if you have a great memory, if you have the interview recorded, you will never need to question whether or not you remember a certain fact correctly. If you do want to make an audio recording of the interview, make sure you ask the witness whether or not they mind if you record them. You don't want to make the witness uncomfortable by recording them against their will. You should also ask them if it's okay to put their real name on your report. You can reassure them though that you will not use their real name or address when you publish the report on a website or any other publication. Most witnesses will not want their personal information published in relation to a Bigfoot sighting. They would likely get an influx of people on their property looking for evidence of a Bigfoot.

7) *Re-ask the same questions*
This may seem like something that you should avoid doing, but it is actually important to do this when conducting a Bigfoot sighting interview. You need to re-ask the same questions because you need to make sure that they are giving the

same answer every time that you ask. There are a lot of people out there who want the notoriety and fame which comes along with the Bigfoot scene. These people might just be making up a story in order to get their fifteen minutes. Other people might give you an answer that they don't really remember the answer to out of fear that you won't believe their story. If this is the case, they often times won't remember the fraudulent answers they gave you previously, especially if it's just a minor detail. This is important because if you ask the same question repeatedly and you get different answers to the question, you can throw out any answer that you got to the question. You can discount those answers as the witness simply doesn't remember without having to bring their change of answer to their attention. If you feel that the answers to the questions change too much, you may want to discount the entire interview as a hoax. This is a point where the only advice we can give you is to use your best judgment. A lot of times legitimate witnesses will be uncomfortable sharing their experience because they feel that you as an interviewer do not believe them. If they seem boastful and still change their answers on some points, they may be simply making the story up.

*Step 3: Have the witness take you to the sighting location*

Despite the fact that you have already spoken to the witness and have obtained all of the details about the sighting, it will still help you immensely to have the witness there with you when you get to the site. The reason for this is that the witness can show you exactly where the sighting took place. Even if you think that you had the right idea from the map that you drew and from the witness' own description of events, it is still entirely possible that you may be investigating in the wrong spot. If the witness is there with you, they can easily show you the exact place where all everything happened.

Here, we would also like to mention that it would be wise to always have another investigator with you. On the one hand, you will not be alone in the secluded wilderness with a stranger at this point in your investigation, so those safety issues will not be a factor. Beyond this, having someone else with you there in the wilderness is some added safety against all of the things that could happen to you when you are out in the wild. Even further, if you have someone with you at the time of the investigation, there is another set of eyes there to see anything that you may have missed.

Once you're there with the witness at the site, have the witness recount their story to you one more time.

*Step 4: Survey the area (with the witness if possible)*

At this point, you, as the investigator, should draw out a more exact map of the area. At this point you should actually measure out distances between landmarks, and you should definitely measure out the distances from the witness to the location where he or she actually saw the creature. This will help you later on when you are making a final report of the investigation. You will have very exact details as to the layout of the sighting area.

Also, at this point you can get a more exact idea of the height of the creature. You can find landmarks that the witness saw the creature walk near, and then you can measure out exact heights. You can show the witness exactly what seven feet looks like and exactly what eight feet looks like. Beyond getting a clearer picture of the creature, the height will help by telling you where the creature may have been able to reach and how high in the foliage you may have to search for hair samples.

*Step 5: Photograph the area*

You can never take too many photographs. Photographs will give you a clearer idea as to the topography and setup of the area. Beyond this, photographs may capture something that you overlook while you are examining the area. Try to capture every angle that you can. Get photographs which show the witness' location and the location of the creature. Get photographs from the point of view of the witness and from the point of view of the creature. Photograph anything that may have been touched by the creature, and photograph anything else that is in the area. Memory cards can hold thousands of pictures nowadays. The tedium of searching through hundreds of pictures is offset by the chance of having photographed something that you missed while in the field.

*Step 6: Search for physical evidence of the creature*

There are many different kind of evidence that you could run across in the field. The best way to search for this evidence, is to first search the immediate area where the creature was seen. Once you have thoroughly searched the path of the creature, search the surrounding area as well for any sign of the creature.

### *Types of evidence to look for*
1) *Footprints*- Look carefully for footprints on the ground where the creature may have walked. It is easier to find these footprints in softer ground, but even harder dirt can sometimes retain a print. Make sure that if you see any possible prints you are careful to not step on it and ruin the impression. Also, make sure that there aren't more prints in the immediate area of the one you found. Also, keep in mind the differences between Bigfoot prints and other prints from known animals (as discussed in the previous chapter.)
2) *Strange impressions in the ground*- Beyond footprints, there are a variety of other impressions in the ground which could be of value to a researcher. Bigfoot handprints

have been discovered, as well as impressions where a creature may have kneeled down in the mud or laid down in the mud. Take notice of any impressions in the ground that you cannot immediately identify.

3) *Hair-* Another item that could be of value in the search for Bigfoot is hair. Hair is often times found on bushes and trees that the creature may have brushed up against. Of course it is impossible to tell if it is Bigfoot hair with the naked eye, but if you find some hair, especially along the same path that the witness saw the creature, it is worth collecting. Samples of hair can sometimes be examined for DNA. Other samples that don't have DNA on them can be examined alongside hair from other known species. Bigfoot hair varies greatly from dear hair and bear hair when examined underneath a microscope.

4) *Broken branches-* Of course there are many things out in nature that can break a branch. Broken branches can also be a sign that a Bigfoot has passed through the area. Of course a simple broken branch is by no means indisputable proof of the creature. At the same time though, if a witness saw a creature in the area, it can suggest that perhaps a Bigfoot has broken the branch. Also, if the broken branches are higher up, this limits the possibilities of creatures that could have snapped the branch. Broken branches can

also suggest that something large has moved through a particular area. While the branches aren't proof of the creature, they are worth documenting because they could give you a clearer picture of the creature you are chasing.

5) *Structures*- The different types of structures that you may run across in the forest will be explained more in depth in a later chapter. There are structures that could be made by wind or natural forces. Other structures are built from branches by known creatures. Still more structures are built by hoaxers in an attempt to fool legitimate Bigfoot researchers. There are strange structures out there that many researchers believe are made by Bigfoot. These structures often times resemble igloos. They are domed structures with an opening to the hollow inside.

6) *Scat*- In our opinion, you should simply leave scat alone when investigating Bigfoot. The first reason for this is that no one, especially us, wants to pick up animal droppings. Secondly, there is not really that much that you can find out from the scat. If you knew what creature the scat came from, you could use it to determine the creature's diet. In this way it would be valuable. On the other hand though, it is impossible to tell what creature left the scat in the field. You cannot determine DNA from scat, so it is impossible to tell whether it was left by a Bigfoot or just

some other creature known to inhabit that particular wilderness. If you want to collect the animal scat, feel free. We just don't see the value in it.

*Step 7: Document your evidence*

In your field kit, you should always have a journal. Every time that you find any piece of evidence that may be Bigfoot related, make sure that you document this find in your journal. Describe it to the best of your ability. Describe where you found it and what the immediate area is like. Beyond this, draw it if you can to give your memory a clearer picture of what you are documenting.

Also, you should photograph any evidence that you come across, no matter how trivial that evidence might seem.

1) *What if I find a footprint?*
   -Look in the immediate area for any other tracks (even known animal tracks for this could prevent you from misidentifying the track of another known creature.)
   -Check for toe impressions. Many things in the wilderness can create impressions in the ground. The easiest way to tell if you have a footprint is to look for the toe impressions. Once you find them, you know that you have a footprint of some sort.
   -Photograph the track. Make sure that when you photograph the track, you put something

in the picture for size reference. The best thing to use for size reference would be a tape measure, but if you don't have a tape measure, you can always use a knife or even your own hand or foot for size reference.

-Measure the track. Make sure that you have detailed measurements of the print's length, width, and depth. The best points to measure depth would be at either the heel or the ball of the foot since these are the areas where the most weight is exerted, so the track will therefore be deepest at those points.

-Measure the distance between the tracks. Make sure that if there are multiple prints, you know how far these prints are from one another. You can determine the stride of the creature from the distance between tracks. Also, make sure you photograph the distance between the tracks, using something for size reference. When you move from one print to another, mark the first print with a stick with a brightly colored flag on it. The reason for this is so that you can clearly tell where the prints are and you do not accidentally step on any of the tracks that you have found.

-Gently remove any debris that is sitting in the impression. Before you make a cast of the print, you need to make sure that the track is clear of any miscellaneous forest debris such as sticks or leaves. Be very careful when removing them though so that you don't disturb the actual impression itself.

Also at this point check for hair within the impression. Footprints are another place where hair is often found.

-Pour your cast. This is the point where you should finally pour your cast to make a permanent record of the print. In the next chapter we will examine in detail exactly how to make an effective cast of a Bigfoot print.

2) *What if I find hair?*

-Photograph the hair. Make sure that you have a record of the environment in which you found the hair.

-Collect the hair. In order to collect the hair, it is best to use either tweezers or latex gloves. The reason for this is to make sure that your own DNA does not contaminate the sample.

-Store the hair in an envelope. Some researchers will store hair samples in plastic bags. We don't think that plastic bags are effective though because the petroleum used in the manufacture of the plastic in the bags could contaminate the sample. Also, plastic bags can trap excessive moisture which can also contaminate the sample. In our opinion, envelopes are the way to go. You want to make sure that the envelope you are using is a plain white envelope. The dye that is used to create colored envelopes could contaminate the sample. Also, make certain that you do

not lick the envelope shut. The DNA from your saliva will contaminate the hair sample.
-Mark the envelope. You will want to mark on the envelope what is stored inside of it. Besides what the sample is, you also want to document the location where it was found, the date on which it was found, and also the time that you found it.

3) *What if I find broken branches?*
-Photograph the branches. Let the photos show the scale of the area. Take some longer shots so that you can tell how far off the ground these branches were.
-Document the height of the broken branches. Make sure that you measure exactly how far off the ground these branches were and how big the creature would have had to have been to break them.
-Check the branches for any further physical evidence such as hair or blood. Hair or blood could belong to anything when checked by themselves in the field. The chances that the hair or blood is from a Bigfoot is much higher when the evidence is connected to an actual broken branch, especially when that branch is too high for most known creatures to have broken.

4) *What if I find a strange structure?*
   -Photograph the structure from all angles. You want to have a comprehensive image of the entire structure for your records.
   -Measure all the sides of the structure.
   -Check the immediate surrounding area for any evidence of Bigfoot activity. If this is indeed a Bigfoot nest, it is more likely that you will be able to find hair or tracks in the area around it.
   -Measure the height and width of the opening if there is an opening.
   -Check carefully for hair around the opening. If Bigfoot are moving into and out of this structure, it is likely that they brush up against the sides of the opening as they are entering and leaving. This will likely cause them to shed some hair, so the opening of these nest structures is perhaps the best place to look for hair.
   -Go inside the structure carefully.
   -Look for physical evidence inside the structure. Again, the structure is likely where Bigfoot spend some time and perhaps sleep, so they could very easily shed some hair within these structures.
   -Look for clues that the structure is man-made. Sometimes, structures such as these are constructed by men. Unfortunately this is done a lot of times to hoax Bigfoot researchers. While you are inside the structure, it is easier to find some of the tell

tale signs that a human built the structure. Look for advanced tool marks and especially keep an eye out for vines that have been tied into square knots. Square knots are a sure sign that the structure was constructed by a human.

-Sniff around for a lingering smell in the structure. A lot of times, Bigfoot will have a musty odor that follows them around. If they have actually spent a fair amount of time in the nest, it is logical to assume that the smell has stayed within the structure somewhat.

-Look for animal or vegetation remains. Many researchers think that these creatures will take many of their meals back to these nest structures in order to eat. Therefore, you should search within these nests for remains of small rodents or some half eaten vegetation. This will support the idea that a creature is living in this structure and will enlighten you as to its diet.

-Measure the interior of the structure. Measure the diameter of the structure and the height of the structure. Determine how many people are going to fit into the structure. Remember that Bigfoot are larger than humans, so if you cannot even fit one person inside the structure, it is likely not a nest for a Bigfoot.

-If you feel that this is an actual nest, set up a hidden trailcam at the location to catch the creature returning to its nest. If you do not have a trailcam, you may want to just stick around and see if any of the creatures return. This is unlikely though because since you have already entered the structure, the creatures are likely to detect your scent and not re-enter the nest.

**What if I don't have a witness, but I still want to go out in the field and search for Bigfoot?**

*Before going out*

Before you go out into the field, it may be best to consult this book or to talk to others about the best possible places in your area to find Bigfoot. If you do not know of anyone you can talk to to find Bigfoot hotspots in your area, you can try to search the local libraries and sift through old newspaper articles or books, looking for historic instances of Bigfoot encounters. Sometimes there are even news stories that are on that speak of Bigfoot sightings in certain localities. These are all possible methods that you can use in finding the place with the highest possibility of yielding Bigfoot evidence.

     When originally researching a site, make sure that you take note of the type of Bigfoot evidence that was discovered in that particular area. If a certain area will yield a lot of footprints, you are probably more likely to find footprints there than any other type of evidence. This doesn't mean that you shouldn't look for other types of evidence, just that you should pay close attention to the type of evidence that is found most frequently at that particular location.

Also, before going out into the field, you should get a detailed map and look at the location. You need to have a very good idea of what kind of terrain you are up against. This map can also be helpful because it should show any major water source that moves through the area. Areas with large amounts of vegetation and a water source are by far the best places to start a search for the elusive creature. Every creature needs to drink.
      Also, you need to make sure that the area that you are going into is large enough to support a creature of Bigfoot's size. There needs to be many acres of wilderness. Bigfoot are not often seen in the suburbs or in any urban area, so the area that supports them needs to be large enough to support them year round and not be inhabited by humans.

Like we mentioned before, it is important for safety and for research to have another researcher there with you at all times. If for any reason you cannot have someone else there with you when you mount your search into the wilderness, make sure that someone else knows exactly where you are going and when you expect to be back. If some accident were to somehow befall you while you are out there, there would then be someone who knows where to start looking for you.

Finally, before you go out into the wilderness, make sure you know what kinds of dangerous wildlife inhabit the area. Know what kind of poisonous snakes or large dangerous mammals such as puma and bears inhabit the area where you are searching. It is easier to deal with these creatures if you know what to expect and how to best avoid them. Beyond that, you know what kind of footprints and evidence you can expect to find out there from known creatures.

*Once you are actually at the location*

Make sure you mark your trail so you can get back out. It is easy to get lost in the woods, especially if the wilderness is vast. GPS units are always helpful in making sure that you don't get lost, but your batteries can always run out. Make sure that you mark the trail that you are using with colorful markers so that you have an easy time of getting back out.

Make sure you go in slowly. Constantly watch your surroundings and listen to the environment. If you hear grunts or movement in the brush, you may be near a creature that you want to pay attention to. Always be on guard, not only because you might miss that crucial piece of evidence, but also for your own safety in the field.

Look for all possible evidence of a large creature. Beyond listening for something large moving in the brush, also keep a close eye out for broken branches and other evidence that something large had recently moved through the area. Also, keep your eyes on the ground, looking for any evidence of footprints or tracks in the dirt.
Look for areas that have large water supplies such as rivers, creeks, ponds, or lakes. These areas are the most

likely to yield evidence of any kind of wildlife in the area, including Bigfoot. Search these areas carefully for any evidence of Bigfoot activity in the area.

Look around for any kind of shelter. This doesn't just mean structures or things that were created by the creature. It also means to look for any caves or other sheltered areas where the creature might sleep and eat in privacy. If you do see a cave, take extreme caution when approaching the cave. If a Bigfoot is cornered, it may attack. And if there is another creature in there such as a bear or mountain lion, it certainly will attack if cornered.

If you happen upon evidence, collect it as mentioned in previous section.

### *In closing:*

Remember to always be skeptical when approaching anything that might be Bigfoot evidence. If you assume that something is Bigfoot evidence before properly examining it, it will remain so in your head despite any evidence you may later come across to the contrary. You need to take a scientific stance when approaching and examining any possible Bigfoot evidence.
      Also remember that your safety is the number one priority. There are many things in the wilderness that can easily kill you. Keep these things in mind and keep a constant and reverent respect for nature and her power. Your own safety comes first, any evidence you come across, no matter how convincing, comes a distant second to that.

# CHAPTER FOUR

# DESCRIPTION OF A BIGFOOT
# BIGFOOT 101

**Bigfoot Description**

Most of the information for this section was adapted from material written by John A. Bindernagel and is used with his permission. Dr. Bindernagel is a wildlife biologist who serves on the TBRC Board of Advisors. Additional regional anecdotal material was provided by the TBRC.

**Body Shape**

Although the initial impression of the sasquatch is its humanlike, rather than bearlike, shape, there are several aspects of its appearance that differ from the human form.

**Large Bulk, Powerful Build, and Thick Chest.**

After registering the initial resemblance to a human, most observers become aware of the large size and overall massiveness of the sasquatch. Although difficult to quantify, the sasquatch's powerful build is reflected especially in its broad shoulders and thick chest. William Roe, describing the female sasquatch he observed within twenty feet of him, noted that "this animal seemed almost round. It was as deep through as it was wide…" Roe went on to say that "we have to get away from the idea of comparing it to a human being as we know them." Similarly, the sasquatch Tom Sewid saw in a British Columbia coastal beach in the spotlight of his fishboat in September, 1994, had a chest "like one-and-a-half forty-five gallon barrels."

**Neck.**

A neck that is absent or is exceedingly short and thick is a consistent field mark, mentioned in many reports. William Roe noted the neck of the one he saw as "'unhuman', being thicker and shorter than any man's I had ever seen." Referring to the adult male he watched for several days, Albert Ostman said that "if the old man were to wear a collar it would have to be at least thirty inches." One witness described the sasquatch he saw as looking like someone in a hooded sweatshirt. Examination of several frames of the Patterson-Gimlin film shows that the apparent absence of a neck in the female sasquatch may be due not only to its hunched posture, but also to the extremely well-developed trapezius muscles which extend from the back of the skull well out onto the shoulders. A young man who came face to face with a sasquatch near Jefferson, Texas in 1989 as a boy reported that "its neck was very thick and maybe short…"

Sketch by Pete Travers of sasquatch seen in 1989 near Jefferson, Texas.

**Face.**

The most obvious feature of the sasquatch face is its relative flatness, clearly lacking the prominent snout of a bear. According to William Roe, the nose of the sasquatch he saw was "broad and flat" but the lips and chin protruded farther than its nose. This slight prognathism (projection of the mouth or jaws) has been reported in other sightings. For example a hunter who watched a sasquatch eat blueberries at the edge of a highway noted it had a humanlike face, "except for the protruding mouth." One female motorist noted that the sasquatch she saw in Newton County, Texas, 1986, had a face that was "flat like [that of] a human…not a pointed nose like a bear."

**Limb Proportions.**

It is frequently noted that, compared to a human, the sasquatch has disproportionately long arms. William Roe, for example, noted that the arms of the sasquatch he saw were "much thicker than a man's arms, and longer, reaching almost to its knees." This is reflected in his daughter's drawing of the animal he observed, done under his direction.

**Hair: Color, Length, and Distribution on Body**

**Hair Color.**

Although sasquatch hair color is most often described as brown (light to dark), black, or simply "dark," other colors have been reported. These include grey, light, white, and silver-tipped. Red (like a Hereford cow) and reddish-brown are also reported with a relatively high frequency. In Texas, Oklahoma, Arkansas and Louisiana, reddish-brown is the second most frequently reported color, after black or dark.

**Hair Length.**

One of the main field marks of the sasquatch is its hairy or fur-covered skin. The hairiness of bears is almost never remarked upon, presumably because hairiness is their natural and expected condition. The presence of hair on the sasquatch, with its humanlike appearance and gait, seems to be more noteworthy. Hair length, as described in numerous sasquatch reports, varies from short to long, and may appear smooth or shaggy. It is often longer on the head, shoulders, and arms than elsewhere on the body.

**Distribution of Hair.**

Patches of bare black skin on the face and chest of sasquatches are sometimes mentioned in reports where close observations have been made. A sasquatch observed near Estacada, Oregon, in 1973 was covered with thick hair except on the throat and chest where the observer saw muscles and skin "the color of a sunburned man." The young man mentioned earlier, near Jefferson, Texas, described the animal he saw at close range as having hair that appeared "dull and coarse." He stated that its hair "covered most of the body, not including the face (with the exception of the cheeks and jawline), forehead, palms of the hands, and the hair appeared much thinner on the thoracic (chest) and abdominal regions." The witness, who had a prolonged daytime view, described the skin as "somewhat tough and weathered," and "medium toned (not black or brown)."

## Anatomical Characteristics of Sex

**Breasts in Females.** The most obvious evidence of gender in sasquatches are the visible breasts of some adult females, such as the one in the Patterson-Gimlin film, and their absence in males. Many observers, such as William Roe, have noted the presence of obvious, engorged breasts and, on this basis, identified the animal they saw as a female.

In most sightings where breasts are observed, they are described as "pendulous" or "long and droopy." The female sasquatch that Albert Ostman watched over a period of several days must have had this appearance, since he stated that "some of those lovable brassieres and uplifts would have been a great improvement on her looks and her figure." In one report, a six-foot-tall sasquatch was described as completely covered with shiny hair except for its face, hands, and the nipple area of its very large breasts. A 2004 sighting report from the Crockett National Forest in Texas had a hunter making a prolonged morning observation of two sasquatches, apparently male and female, while they were "turning over fallen limbs and eating something from under them." The hunter said of the female, "'she' had pronounced breasts…" A fisherman in Sebastian County, Arkansas, described the sasquatch he saw in 1988 as having "what looked like large floppy breasts."

## Differences From Human Gait

The sasquatch bipedal gait differs not only from the quadrupedal gait of other mammals but also from the human bipedal gait in several small but significant ways. Many observers have remarked on the gracefulness of the sasquatch stride. One man likened the sasquatch he saw to "a really fit athlete" as it strode off. A woman who watched a sasquatch stride across a beach on the west coast of Vancouver Island in August, 1995 said it moved "as if it was walking on air," and an Ohio observer watched a sasquatch cross a field "taking strides as if it were walking on eggshells." One Texas woman, while in the Sam Houston National Forest, in Walker County, Texas, 2005, described a sasquatch she observed as moving "to the side of the road in a smooth motion." She went on to say that its movement was "fluid—like it was floating." Similarly a man who watched a sasquatch near the Alaska Highway in the Yukon noted that its motion was smooth and "there was no bobbing movement at all."

One may wonder just what it is about the sasquatch gait that renders it so graceful. The Patterson-Gimlin film, despite its lack of visual detail, clearly shows the smooth, ground-eating strides of an adult sasquatch walking along a sandbar. Several investigators, including professor Grover Krantz of the University of Washington, have examined this film at length. The results of his investigations show that the sasquatch has a bent-knee gait wherein the knee does not lock during the stride as does the knee of humans. This bent-knee stride, which we humans employ in the diagonal stride of cross-country skiing and in fast ice-skating, lends fluidity and removes the vertical motion inherent in the normal human stride. At least one man who watched a sasquatch walk away with a very smooth, rapid gait noticed this "peculiar" posture and observed that "its legs were always bent as if in a slouch." A woman in Rogers County, Oklahoma, in 2002 described a sasquatch that she observed as "somewhat stooped; it slowly swung its arms and had a slight knee bend."

The arm-swing of the walking sasquatch is also noteworthy. First, it is worth noting because it is present at all, and second, because it sometimes appears exaggerated when compared with the normal walking gait of humans. In the Patterson-Gimlin film, for example, the hand is raised high in front of the body, describing a longer arc than humans normally exhibit. In 2007, a security guard who reported a late night visual encounter with a sasquatch near The Woodlands, in Montgomery County, Texas, remarked that it had an "odd arm swing," and "had a very long unusual stride." Todd Neiss, describing the three sasquatches he observed near Seaside, Oregon, in April 1993 remarked on the "exaggerated arm movements" of the animals as they walked swiftly out of sight.

A less obvious difference is the tendency of the sasquatch to pick up its feet at the end of a stride to a greater extent than humans normally do. Like the arm-swing, this action is an exaggeration of a normal human action, and brings the sole of the foot to an almost vertical position at the end of each stride.

**Occasional Quadrupedal Gait**
In the spring of 1996 an observer near Cambridge, Ohio, saw an adult sasquatch with a four-foot-tall juvenile. The juvenile was "bouncing around like a chimpanzee" and "never did walk totally erect like the other one." The situation in infant sasquatches may be similar to that in humans in which young individuals take many months to "learn to walk" in the sense of mastering bipedal locomotion.

Although bipedalism appears to be the norm in adult sasquatches, there are a few reports where quadrupedal locomotion was observed. The first, a historical report (1869) of a "wildman or a gorilla or 'what-is-it'" from eastern Kansas and western Missouri, notes that the animal "generally walks on its hind legs but sometimes on all fours." A more recent report from western Florida in 1975 describes two boys being chased by a "skunk ape that smelled terrible and ran on all fours in a crablike fashion as well as running upright." A witness who, as a boy, observed a sasquatch at close range in 1989 near Jefferson, in Marion County, Texas, remarked how the animal he saw was emerging from dense woods "moving on all fours" until it hoisted itself up using a nearby fence-post, transitioning smoothly into a bipedal posture and gait. The young man noted that the "change in gait or posture did not result in a change in its speed." When the animal fled, it "took off sprinting on two legs, using his hands to tunnel his way through the thicket."

## Upright Stance and Hunched Posture

It should be noted that although sasquatches stand and walk fully upright on two legs, they often assume a hunched posture.

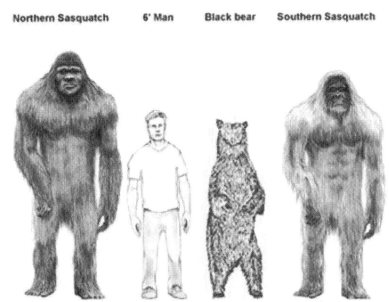

Comparison sketches of sasquatches, a man, and black bear.
Drawings by Pete Travers and Wendy Dyck.

The hunched or stooped posture noted in many sasquatch reports may have an anatomical basis. A.F. Dixson shows that whereas in humans the spine enters the base of the skull, in the gorilla it enters the rear of the skull. This configuration may be partly responsible for the apparent absence of a neck in the sasquatch in front or rear view. It also restricts turning of the head so that the entire upper body must rotate in order for the animal to look sideways or rearward. This restricted head movement is visible in the Patterson-Gimlin film and has been discussed by Grover Krantz in Big Footprints.

When standing, sasquatches may slouch with their long arms dangling. Occasionally they lean sideways against a tree using an extended arm for support.

**Crouching, Squatting and Sitting**

**Crouching.**

Sasquatches typically move about in a crouch or stooped position when foraging. For this reason they are frequently mistaken for bears until they stand erect or walk away bipedally. One example of this occurred in the spring of 1997 in British Columbia's Fraser Canyon where a bear hunter watched what he thought was a bear through his rifle scope. While he was waiting for the "bear" to turn sideways so he could assess its size, the animal stood up on its hind feet, reached above its head, and with its hand pulled down the tip of a tree branch from which it ate the leaf buds. It took a step forward and repeated this process, still on its hind legs. By the time the animal stood erect the hunter realized he was not watching a bear, but a completely different animal. (In addition to its wide shoulders he observed bare skin on its flat face and on the sides of the chest below its arms.)

**Squatting and Sitting.**

When sasquatches pause or rest they may squat, or sit on a stump or similar support. Squatting sasquatches have also been reported in shallow water where they were apparently harvesting water plants with their hands, or on land, where they have been seen digging or handling roots with their hands.

**Swimming**

Swimming must be examined alongside the terrestrial gait of the sasquatch since it appears to be an important means of locomotion throughout the range of this species in North America, especially on the west coast. Circumstantial evidence, such as reports of the presence of sasquatches on small islands off the coast of British Columbia, has suggested they swim. Observations of sasquatches actually swimming have confirmed this.

In a 1965 sighting on the northern British Columbia coast, a boatman saw a sasquatch swimming towards a rocky islet on which two others were standing. "The one in the water swam very powerfully and very fast, with the water surging around its chest."

There are at least two reports of sasquatches swimming in large rivers. In Montana a sasquatch was reported swimming from the west bank of the Missouri River to Taylor Island in August, 1976. And in Grand Rapids, Manitoba, nine members of a family saw a sasquatch swim up the Saskatchewan River past their house and then come out and walk into the bush a quarter mile away.

In his book In the Big Thicket: On the Trail of the Wildman, Rob Riggs wrote of "Old Mossyback," reportedly an "ape-like creature" that dwelt in and around the Trinity River of Southeast Texas. Riggs told of one particular man he interviewed named John who presumably saw "Old Mossyback" as it raided his rabbit pen and made off with one of the rabbits near Dayton, Texas. John pursued the "large, dark form," staying just close enough to hear the terrified rabbit's screams. John made it to the bank of the Trinity River and watched in the moonlight "a huge ape-like animal" as it swam to the other side of the river still holding onto the rabbit.
A Washington couple, fishing in the Nooksack River in September, 1970, saw a black eight-foot-tall, slightly-stooped animal with a flat face and no neck standing in water up to its knees about two hundred yards away. It bent down and disappeared in the muddy water. Later, tracks were found coming out of the water onto a sandbar for about 150 yards and then back into the river. (The tracks were thirteen-and-a-half inches long with a forty-five inch stride.) The wife saw the sasquatch again later in the month when it stood up at the front of her boat in water about five feet deep which came only to

the top of its legs. These sightings took place during a heavy salmon run; other sasquatch sightings were made during the same period in the same area.

South of Lummi, Washington, not far from the above location, a young girl watched while a sasquatch stood and swam off the beach in front of her house at dawn as fog was lifting. It seemed to be fishing.

Finally, from Ketchican, Alaska, there is a report of a fifteen-year-old boy who saw a humanlike figure standing waist-deep in the water between the float of a dock and the shore. When he screamed and fled, about thirty men came out of shock on the dock and saw the animal as it dove under water and swam away. Looking down they could see it swimming underwater with its arms forward and legs doing a frog-kick until it swam out of sight. Prior to this, these fishermen had been troubled with something ripping nets and stealing fish. This report is significant because it describes the stroke used by at least some sasquatches when they swim.

Several of these reports suggest that swimming in sasquatches not only serves as a means of locomotion for travel, but may be sufficiently developed to serve in catching fish.

# Eastern Bigfoot/OhioGrassman
## Height: 7-8 ft.
## Weight: 300-500 lbs.

# Pacific Northwest/Canada
## Height: 8-10 ft.
## Weight: 400-600 lbs.

# North America/Canada
### Height: 8-12 ft.
### Weight: 600-1000 lbs.

# Southern/Florida Skunkape

### Height: 6-8 ft.
### Weight: 300-500 lbs.

## Southern/Texas Arkansas
### Height: 7-8 ft.
### Weight: 300-500 lbs.

# Australian Yowie

### Height: 7-8 ft.
### Weight: 300-500 lbs.

# THE YOWIE

**The Yowie** (also called **Yahoo** or the **Great Hair Moehau** in New Zealand) is an alleged hominid reputed to lurk in the Australian wilderness, similar to the Himalayan Yeti and the North American Bigfoot.

## Description

The Yowie is very similar to Bigfoot with a height of six to seven feet tall and a thick black or brown fear covering the entire body. He is bipedal but has been seen running on all four legs at times.

## Habitat

The Yowie has been reported primarily in New South Wales, the Gold Coast of Queensland and in the wild bush country of the Moehau Range. In New Zealand, the North Auckland area and the West Coast are its favorite playground.

## Sightings

The first recorded sighting of a Yahoo by a European came in 1881, when an Australian newspaper reported that several witnesses had seen a large baboon-like animal that stood taller than a man. In 1894, another individual claimed to come face to face with a "wild man or gorilla" in New South Wales bush. A 1903 newspaper printed the testimony of a man who said he watched as aborigines killed a Yahoo, which he said looked "like a black man, but covered all over with gray hair."

In 1912, George Summerell was riding on horseback

between Bombala and Bemboka when he saw a strange creature on all fours drinking from a creek. The animal rose up on its hind feet to a height of seven feet and looked at Summerell. Then it disregarded the horseman, finished its drink, and peacefully walked away into nearby woods. The following day, Summerell's friend Sydney Wheeler Jephcott rushed to the scene of the sighting and discovered an abundance of handprints and footprints. Jephcott described the footprints as humanlike but huge, and having only four toes per foot. He said he made plaster casts of the tracks and turned them in to a local university, but there is no record of a scientific analysis being rendered. In 1971, a Royal Australian Air Force helicopter carrying a crew of surveyors landed atop Sentinel Mountain, a remote and inaccessible peak. Much to their surprise, the team discovered fresh footprints in mud, much larger than human footprints, in a place where no known biped could possibly be present. Yowie sightings continued steadily throughout the '70s. In 1976, backpackers in New South Wales reported seeing a five-foot female Yowie whose fur stank to high heaven. Also in New South Wales, Betty Gee reported seeing a giant creature covered with black fur outside her home in 1977. Shortly thereafter, her fence was knocked down and large black fur outside her home in 1977. Shortly thereafter, her fence was knocked down and large footprints surrounded the scene. A man in the Gold Coast city of Springbrook said that a"big black hairy man-thing" appeared before him while he while chopping wood in 1978. "It just stared at me and I stared back," he said. "I was so numb, I couldn't even raise the axe I had in my hand."

In 1997, a woman residing in Tanimi Desert was awakened at 3 a.m. by a horrible animal-like noise just outside her bedroom window. When she went out to investigate, she was confronted with an unbearable stench that sent her into the dry heaves, and she saw a seven-foot hairy creature tear through her fence as it made a hasty retreat. The next day, police discovered a number of giant ootprints and a somewhat shredded irrigation pipe that had seemingly been chewed upon.

## Origins

Jonathan Swift's *Gulliver's Travels* (1726) includes a subhuman race called the Yahoos. Hearing the aborigines' fearful accounts of this malevolent beast, nineteenth-century European settlers probably applied the name Yahoo to the Australian creature themselves. Sometime in the 1970s, the term "Yowie" supplanted "Yahoo," for reasons that remain as mysterious as the creature. One possible origin of the newer name is the aborigine word **youree**, described as a legitimate native term for the hairy man-monster. The Australian accent could easily contort "youree" into "Yowie."

The earliest published reference to the word in its current usage is in Donald Friend's *Hillendiana,* a collection of writing about the goldfields near Hill End in New South Wales. Friend refers to the "Yowie" as a species of "bunyip", an Aboriginal term used to describe monsters said to dwell in many Australian rivers and lakes. Paranormal enthusiast Rex Gilroy popularized the word in newspaper articles during the 1970s and 1980s

## Folklore

An aborigine folk tale explains that when their people first migrated to Australia thousands of years ago, they encountered on the new continent a savage race of ape-men. The aborigines' ancestors went to war against the ape-men, and in the end the humans triumphed, thanks to their ability to make weapons. Some have wondered if this tale might contain some element of truth, and it is a few diehard survivors from this unknown primate species that would later be known as the Yahoo and the Yowie.

## Theories

Australian Rex Gilroy, a self-proclaimed archeo-cryptozoologist, has attempted to popularize the scientific term *Gigantopithecus australis* for the yowie in his book *Uru - The Lost Civilisation of Australia*. He claims to have collected over 3000 reports of them and proposed that they comprise a relict population of extinct ape or Homo species. There is, however, no evidence that Gigantopithecus ever existed in Australia.

# **The Yeti**
### Height: 6-8 ft.
### Weight: 300-500 lbs.

# The Yeti

The **Yeti** or **Abominable Snowman** is an alleged ape-like bipedal cryptid said to inhabit the Himalayan region of Nepal and Tibet.

Some in the scientific community regard the Yeti as a legend, however it remains one of the most discussed and seeked creatures of cryptozoology, together with his "cousin", Bigfoot of North America.

## Etymology and names

In Tibetan, Yeti means "magical creature"; meh-the means the "manlike thing that is not a man". The appellation "Abominable Snowman" was not coined until 1921, when Lieutenant-Colonel Charles Howard-Bury led the joint Alpine Club and Royal Geographical Society "Everest Reconnaissance Expedition" which he chronicled in *Mount Everest The Reconnaissance*, 1921. In the book, Howard-Bury includes an account of crossing the "Lhakpa-la" at 21,000 ft (6,400 m) where he found footprints that he believed "were probably caused by a large 'loping' grey wolf, which in the soft snow formed double tracks rather like a those of a bare-footed man"

He adds that his Sherpa guides "at once volunteered that the tracks must be that of "The Wild Man of the Snows". However, the translator at the time incorrectly transcribed it as *metoh-kangmi*, which translates approximately to "abominable snowman".

Other terms used by Himalayan peoples do not translate exactly the same, but refer to legendary and indigenous wildlife:

- **Meh-teh** translates as "man-bear".
- **Dzu-teh** translates as "cattle bear"
- **Migoi** or **Mi-go**) translates as "wild man".
- **Mirka** - another name for "wild-man", however as local legend has it "anyone who sees one dies or is killed".
- **Kang Admi** - "Snow Man".
- The people of Nepal also call them **rakshasa** which is Sanskrit for "demon".
- **Jo-Bran** - "Man eater".

## Foklore

According to the Tibetans, stories of its existence date back to the 4th century BC when references to the Yeti are found in a poem called 'Rama and Sita'. The belief in these creatures is universal among Tibetans. The government of Nepal from officially declared the Yeti to exist in 1961. It became their national symbol, and an important source of income.

Some relics are displayed in Tibetan monasteries such as a so-called yetit scalp

## Habitat

Most sightings of the Yeti have been recorded in Mongolia, Tibet, Nepal, Pakistan, and surrounding areas. The Himalayas, which include the famous Mount Everest (the highest mountain in the world with 29,028 feet high) offer many areas that are not accessible to man and hidden caves where large population of the creature are said to inhabit

## Species

According to legends, there are three species which have a general resemblance despite differences in size.

## Teh-Ima

**Aka :** pygmy yeti, thelma, Raksi-Bombo

**Description:** Diog thinks he is probably a gibbon (a known type of ape) that may live as far north as Nepal, though it's never been spotted past the Brahmaputra River in India.

**Size :** 1.0 -1.5 m in height

## Dzu-teh

**Aka:** giant yeti, Nyalmot

**Place :** Nepal, North Vietnam and Sikkim

**Description:**. The giant yeti has dark fur and a flat head and has been said to roam and often attack cattle. Diog thinks this is probably the Tibetan blue bear. A creature so rare it is known only in the west through a few skins, bones and a skull

**Size :**2.5 to4.5 m in height

## Meh-teh
**Aka** : Mih the, Rimi
**Description:** see above description
**Size :**2 to 3 m in height

# The Russian Bigfoot
# Height: 7-10ft.
# Weight: 400-500lbs

# THE RUSSIAN BIGFOOT
# (ALMAS ALMASTI )

Alma

Alma n : a creature reported to be of ape-like appearance that inhabits the mountains in central Asia, which was up until a few years ago part of the Soviet Union.

Although not as well known as the Yeti and Bigfoot stories about the Alma suggest that it is a creature more akin to a hairy human than an ape.

Professor Boris Porchnev, of the Moscow Academy of Sciences, published a description of the creature based on detailed stories he'd gathered from people who had seen it.

"There is no under layer of hair so that the skin can sometimes be seen," says the report. "The head rises to a cone-shaped peak," it continues, and "the teeth are like a man's, but larger, with the canines more widely separated." Porchnev's description also noted that the Alma can swim in swift currents and run as fast as a horse. Breeding pairs remain together living in holes in the ground. Their diet consists of small animals and vegetables, and they have a mainly nocturnal nature. It is also noted in the report that Almas have a "distasteful smell."

A traveler in Mongolia called N.M. Pzewalski gathered the first stories of the Alma in 1881. He was also responsible for the discovery of the Mongolian wild horse. Sightings of the Alma were reported during the Second World War by refugees, soldiers, and prisoners of war.

There have been reports that Almas have been shot and killed. One such report happened during a clash with the Japanese by a Russian reconnaissance unit in Mongolia. Two shadowy figures were shot when they failed to respond to a challenge by sentries. Unfortunately because of the war the bodies, described as having the appearance of a "strange anthropoid ape" covered with long red hair and about the size of a man, could not be returned to Moscow for thorough scientific evaluation.

Several expeditions were undertaken to look for the Alma while the Soviet Union was still together. However, although many interesting stories turned up, there was little in the way of unusual artifacts. Lack of hard evidence led Russian scientists to become disillusioned about the value of searching for Almas.

Dr. Proshnev did speculate that the Alma could perhaps be one of a last surviving group of Neanderthal men. Indeed the areas where reports of Alma sightings have taken place have yielded a number of Neanderthal artifacts, thus lending some weight to Proshnev's hypothesis.

Political instablity and the dismemberment of the Soviet union has done little to facilitate further research in this area, so for the time being the Alma remains an enigma.

**The Orang Pendek**
**Height: 4-6 ft.**
**Weight: 200-300 lbs.**

# THE ORANG PENDEK

**Orang Pendek** ( Indonesian for "short person") is the most common name given to a cryptid, or cryptozoological animal, that reportedly inhabits remote, mountainous forests on the island of Sumatra.

**Aka** : Batutut, Atu, Sedapa, Mawa (Malay).

**Folklore** : Natives of Sumatra have generally accepted the Orang-Pendek as a genuine animal for centuries, and because they believe it to be a gentle creature that only attacks small animals for food, they regard it with tolerance and respect, rather than fear.

**Place**: Sumatra (the Taman National Kerinci Seblat and bordering forest), the Sarawak/Kalimantan border in central Borneo, and presumably other islands in Indonesia and Malaysia.

The animal has allegedly been seen and documented for at least one hundred years by forest tribes, local villagers, Dutch colonists, and Western scientists and travelers.

Consensus among witnesses is that the animal is a ground-dwelling, bipedal primate standing about two and a half to five feet tall, with human and ape characteristics evidently not lacking in strength, speed and agility.

The creature is said to have a pinkish-brown skin

covered by a short, dark fur with a mane of long hair around the face that flows down the back.

The Orang-Pendek is said to walk mostly upright and to possess relatively short arms. It is said to walk upright at incredible speeds, does not brachiate, although it has been seen sitting in or hugging trees. Pint-sized footprints about six inches long, shaped very much like human footprints except for being proportionately rather broad and having a divergent toe, have been presented as evidence of the creature.

Debbie Martyr, a prominent Orang Pendek researcher who has worked in the area for over 15 years, has interviewed hundreds of witnesses, and alleges to have seen the animal personally on several occasions. ...usually no more than 85 or 90cm in height — although occasionally as large as 1m 20cm. The body is covered in a coat of dark grey or black flecked with grey hair. But it is the sheer physical power of the orang pendek that most impresses the Kerinci villagers. They speak in awe, of its broad shoulders, huge chest and upper abdomen and powerful aims [sic]. The animal is so strong, the villagers would whisper that it can uproot small trees and even break rattan vines. The legs, in comparison, are short and slim, the feet neat and small, usually turned out at an angle of up to 45 degrees. The head slopes back to a distinct crest — similar to the gorilla — and there appears to be a bony ridge above the eyes. But the mouth is small and neat, the eyes are set wide apart and the nose is distinctly humanoid. When frightened, the animal exposes its teeth — revealing oddly broad incisors and prominent, long canine teeth.

The Orang Pendek - Theories

Orang Pendek's reported physical characteristics differentiate it from any other species of animal known to inhabit the area. All witnesses describe it as an ape- or human-like animal. Its bipedality, fur coloring, and southerly location on the island make orangutans an unlikely explanation, and its bipedality, size, and other physical characteristics make gibbons, the only apes known to inhabit the area, unlikely as well. Many therefore propose that Orang Pendek could represent a new genus of primate or a new species or subspecies of orangutan or gibbon.

As far back as Mr. Van Heerwarden's account of Orang Pendek, people have speculated that the animal may in fact be a "missing link" (a hominid representing an earlier stage in human evolution). In October 2004, scientists published claims of the discovery of skeletal remains of a new species of human (Homo floresiensis) in caves on Flores Island (another island in the Indonesian archipelago) dating from 12,000 years ago. The species was described as being roughly one meter tall. The recency of Homo floresiensis' continued existence and the similarities between its physical description and the accounts of Orang Pendek have led to renewed speculation in this respect.

Other powerfully built, hairy, man-like creatures seem to have lived near Laotian/Vietnam border until the bombing of the Ho Chi Minh Trail destroyed their habitat. They were called **Nguoi-rung** by the locals.

Sightings seem to be concentrated on the hills of the Chu Mo Ry near the Cambodian border.

**The Chinese Yeren**
**Height: 7-9 ft.**
**Weight: 600-800 lbs.**

# THE CHINESE YEREN

**Wild**The **Yeren** (Chinese: 野人; literally "wild-man"), also known as the **Yiren, Yeh Ren, Chinese man**, or **Man-Monkey** is an alleged hominid residing in the mountainous forested regions of China's remote Hubei province.

## Description

Wang Tselin describes him as follows

> "The creature stands fully erect like a human an are about 6 feet tall. The legs are of human proportions relative to their stature with elongated arms. The level of the chin is above that of the shoulder, a very human trait but there is little projection of the nose. Interestingly, the slopingforehead rise up above the eyes like in humans rather than back like in the gorilla. It has deep-set eyes, protruding lips, horse-like front teeth and bulbous nose with slightly upturned nostrils. He has sunken cheeks, ears like a man's but bigger, and round eyes also bigger than a man's. His eyes are black and he is covered in long, dark brown. His whole face, except for the nose and ears, is covered with short hairs. His arms hang down to below his knees. He has big hands about half a foot long and with thumbs only slightly separated from the fingers. He walks upright with his legs apart. His feet are about a foot long."

**Size** :Its height varies from 6 to 9 ft tall. A smaller

version that is only 3 ft tall has also been sighted. hairy all over. It had thick lips and big teeth, like a horse's. and its arms were very long and its feet were huge.

**Color** :Reddish or brownish (sometimes greyish) hair covers his body

## Folkore

The Wildman has been a part of the folklore of southern and central China for centuries. Apparently ancient Chinese literary works and folk legends include references to big hairy manlike creatures which live in the vast forests of the Quinling-Bashan-Shennongjia, a mountain region of central China (northwest Hubei province).

It seems that even Shennong, god-king of fable and father of husbandry and farming, was deterred by their altitude and had to build a scaffolding when he came here to collect medicinal herbs. The area even got its name from Shennong and the Chinese word *jia* meaning 'scaffold'.

Two thousand years ago, during the Warring States period, Qu Yuan (340-278 B C), the statesman-poet of the State of Chu, referred in his verse to 'mountain ogres'. His home was, significantly, just south of Shennongjia, in what is today the Zigui county of Hubei province.

In Tang dynasty times (A D 618-907) the historian Li Yanshou in his Southern Historydescribes a band of 'hairy men' in the region of modern Jiangling county, also in Hubei.

Later still, the Ch'ing dynasty poet, Yuan Mei (1716-98), in his book *New Rhythms* tells of the existence of a creature described as 'monkey-like, yet not a monkey' in south-western Shaanxi province, Xianning county. The first Chinese emperor, Hwang-Ti, builder of the Great Wall, may have had an unwitting hand in Yeti-making. According to an ancient legend, some people tried to avoid compulsory labour on the wall by taking to the forests and hiding there where, even after many generations, their descendants became wild, large and hairy but retained the power of speech. They emerged periodically from the forest and enquired, 'Has the wall been finished yet?'. But, although the answer was 'Yes', they didn't believe it and returned to the forest where, alas, reality is about to catch up with them

## Theories

The Chinese incline towards the view that their creature is related to

*Gigantopithecus*

, a giant extinct primate believed to have lived in China three hundred thousand years ago.

A monkey species that has also been suggested as a candidate for Wildman sightings is the rare and endangered *golden monkey*, whose unusual appearance could seem like a man-monster to some observers.

# CHAPTER FIVE
## By
### The American Bigfoot Society
### Track Casting

If you are considering field research, one of the first issues you must address is "to cast or not to cast."
This is a decision that should be made prior to any field work, so you may obtain the necessary casting materials. My personal field bag for casting always contains these items:

1. Plaster of Paris
2. 80 oz bucket (for mixing)
3. Water
4. Latex Gloves (to be thrown away after use)
5. 6 ½" Aluminum Tie Wire 11GA
6. Small baster. (for water removal from track, if any)
7. Tape Measure (Metal)
8. Small note pad (to write down the various measurements and any information worthy of note that you do not want to forget)
9. Tweezers

From all descriptions, Bigfoot may very well be a primate. All primates have dermal ridges and flexion creases. What are dermal ridges and flexion creases? Dermal ridges are your finger prints, flexion creases are the deep lines of the hand. These are unique to every person and non-human primate. It has been my experience through years of experimentation with casting, that dermal ridges and flexion creases can be captured during the process of casting.

First thing, first; Take photos of the track, in the soil. Take the photos low as sunlight or the camera flash can wash out whatever it is you are trying to capture in the photo. Measure the cast, and take photos of the cast along with the tape measure showing the length. This is a good resource for later. Record measurements from the toe to heel and from the left side of the instep to the right. Examine the track for toe impressions, or anything of interest. If you see anything within the track that resembles dermal ridges or flexion creases, photograph them as well.

Once the examination of the track is complete and all documentation has been performed, now it's time to cast that track.

I use Plaster of Paris. Plaster of Paris captures the details of a track with just as much accuracy as Hydrocal B-11 or Ultracal 30. Plaster of Paris is less expensive and a copy of the track can be made later using Hydrocal B-11 or Ultracal 30 if the finished cast is worthy of keeping forever should it show signs of dermal ridges or flexion creases.

The most important bit of information I can give is simply this. Follow the manufacturer's instructions on the back of your bag or box of Plaster of Paris. There are good reasons why these instructions are there.

**<u>Mixing Instructions</u>:**

Plaster of Paris is a two to one ratio. Meaning:

2 cups Plaster of Paris
1 cup Water

Pull out your latex gloves, and put them on. Fill your bucket with 2 cups water, first. Then once you have the water in your bucket, measure out 2 cups of Plaster of Paris. Slowly add the Plaster of Paris to the water, allowing the water to take it in. Do not force this by shaking the bucket or forcing it down by hand. Allow this process to take place. Doing so will allow the water to force out the air that is trapped within the Plaster of Paris. Trapped air can lead to what is called, "Casting Artifacts." While the water is taking in the Plaster of Paris, continue to slowly add more until you have put in the entire 2 cups of Plaster of Paris. You will notice how the Plaster of Paris "piles up" in the bucket, this should happen. Continue to allow the water to take in the Plaster of Paris until it turns from a "white" to "grey". After all the Plaster of Paris is in the water, and has turned "grey" you may now use your latex gloved hand to stir the mixture Using your hand will guard against more air becoming trapped during the mixing process and also allows you to identify (by feel) any clumps that may be inside the mixture. Be sure to break up these clumps and mix until it is creamy and smooth. If you must use something, other than your latex gloved hand to stir, make sure whatever you use is not smaller than the width of your hand.
A word of caution, once you have begun the process of adding the Plaster of Paris to the water, you must continue until the Plaster of Paris is poured into the track. Once the casting agent hits the water, it begins a

process called, "kicking." If you take too long, the Plaster of Paris can set up in your bucket, forcing you to start all over again.

**Pouring the Cast:**
Carefully remove any forest litter that can be easily removed with a pair of tweezers. If you must pull anything out of the soil, leave it alone. Better to leave it, than have the track ruined due to the soil being pulled up.
Once you have the casting agent in the bucket of water and mixed to a smooth creamy consistency you may now begin the process of casting the track. I use a method called "splash casting". To do this, you simply put some of the mixed Plaster of Paris on your fingertips and "fling" the plaster into the bottom of the track and the sides. "Flinging" the Plaster of Paris causes it to stick where it hits. It will not run and wipe out any details you may be trying to capture. Once you have the bottom of the track covered using this method you can simply pour the remaining casting mixture until the track is completely filled. If you are uncomfortable with this method, simply pour the mixed Plaster of Paris into your hand, allowing it to run off your fingers tips, and onto the desired areas of your cast. Use caution while pouring the Plaster of Paris and take your time so the plaster hitting the bottom of the track does not wipe out any detail that may be there.

**Removing the Cast:**
Once your cast is poured I would wait no less than forty-five minutes, before attempting to remove the cast from the soil. If the top of the cast has a shiny look, it is still in the kicking process and should be allowed to sit. You

will notice differences in set time depending on soil conditions (dry vs. wet soil) and humidity levels. If the soil you casted in is wet, allow more time. If you attempt to pull the cast too soon, you risk breaking and destroying the cast. If you can, take the extra time, I would highly recommend it.

During this waiting time, I dig a "moat" completely around the track, in a large circle. Doing this will allow you to cut under the cast (when dry enough) and remove the entire track, dirt and all. I then place the cast and dirt on a dry surface, (cardboard works well) dirt side up and get it to a dry place where it can finish setting up. Allow your cast to sit, undisturbed, for at least 24 hours before attempting to clean.

**Fixatives:**
There is much discussion in reference to "fixatives" (hairspray, etc.) to be sprayed in a track prior to casting. Personally, I do not use them. During my own casting experiments I have successfully casted my own dermal ridges and flexion creases, without the aid of a "fixative." In more cases than not, these sprays have wiped out the fine details of the dermal ridges I intended to cast of my own foot. My motto is, "If it's not necessary, why do it?"

**6 ½" Aluminum Tie Wire 11GA:**
What is this used for? I use these for helping to strengthen the cast. These tie wires are made of Aluminum, so they are rust resistant. My suggestion is to fill the track to about half, then stop and carefully lay two-three in the Plaster of Paris (being very careful not to allow these tie wires to fall to the bottom) then once you have these laid in, simply finish pouring the rest of

the track to the top. I usually make enough Plaster of Paris to cover all the way to the top, and approximately one-two inches around the outside of the track. Doing this will help show the depth and finished detail of the track.

**Small Baster:**
Why bring a jackhammer when a hammer works? A smaller version of a baster, I have found, is much easier to use when trying to pull water from a track. I have used the full size versions, but have found they have quite a bit of force, which means the water being removed stirs up the dirt in the bottom of the track, thereby causing you to lose any detail you may have and not even know is there. You can purchase a small baster at any grocery store, and most often they are under five dollars.

**Oven Drying Your Cast:**
There are some who stick their casts into the oven. Beware; if you set the oven temperature too high, your cast will break and that will be that. Allow the cast to dry in its own time. You will know it's dry and ready for cleaning when you touch the Plaster of Paris and it is dry to the touch.

I always recommend people take some time and practice, before the big moment comes. Not every soil type will allow for fine details like dermal ridges, the sandier or rockier the soil the less your chances will be for capturing fine details.

By practicing you allow yourself to get used to the materials and the soil. You will be far more confident and that confidence will allow you to feel good about the outcome.

We cast for the potential evidence that may be contained in the track. Take your time, be patient and you will be happy with the outcome, and you will have results you can rely on.

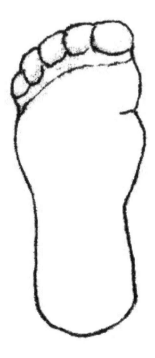

# CHAPTER SIX

<u>Tristate Bigfoot – Investigation Techniques and Procedures</u>
www.tristatebigfoot.com

Setting up basecamp
When we first arrive to our basecamp location the first thing we do is perform a parameter check. We check the parameter for any evidence or signs of wildlife that may frequent the area. Checking for scat, and animal tracks. If you know what animals frequent the area it will come in handy in identifying different sounds or noises that you may hear or pick up on your audio recordings. We also check for any broken trees or limbs and document these by taking notes of their location and such. This will help you determine if the tree breaks found are old or happened during the investigation. We also check and make note of any edible vegetation while doing the parameter check.
Setting up equipment and determining where the equipment will be placed should be your first priority after performing the parameter check. The setting up of any bait stations or any other methods of attempting to lure a Bigfoot should be put in place at this time also.If you own a GPS device, marking the placement of your equipment with this device may aid you in retrieving it later. Marking the basecamp location with your GPS unit is also recommended.

After base camp has been set up, including the sleeping and eating areas of the researchers, a quick map should be drawn up to record the placement of the equipment and the entire camp for your records. If you ever plan on returning to this area, it may be helpful later to know what was placed where and if there was anything noteworthy. Make notes of all the important data such as weather conditions, air temperature, moon phase, etc. Also determine and make note of the nearest water source. It is recommended that at least one sound recording device is running at all times during the daytime hours. After dark we recommend at least two. It is helpful to really know your equipment. Know the recording length, battery life, and extra power supplies on hand. Recording equipment should be started at different times, staggering them so that there is at least one device running while other devices are being prepared. Having a schedule of which device will be used and when is very helpful.

Make sure that you label any tapes with the date & times. If you use digital devices make sure to speak the dates and time into the recording devices, especially voice recorders. Speaking the time at certain intervals is also recommending while recording.

Investigating the area where an encounter took place. If you experience an encounter, have a sighting report, or if you are checking out an area where a sighting was reported; one of the first things that you need to do is collect the data for this location. Record the coordinates and elevation. A detailed description should also be made of this area, the distance to the nearest house or road and nearest water resource, etc. Take note of the area's plausible food sources. Record the date and time the event took place. Make note of the air temperature and weather conditions at the time of time of the encounter if possible. Be descriptive as possible of the terrain. Note evidence of other animals that are in the area.

If you are there with the witness or you witnessed it yourself, video tape the area while you or the witness is describing the event that took place. Take pictures and make casts of any footprints or other evidence that is found.

Collecting Physical Evidence
Photograph the sample from every angle before you collect it. Also photograph the surrounding area. If you find multiple samples it will be helpful to number them. Place a piece of paper with the corresponding number in any pictures taken of that sample and its surroundings. It is important to label any evidence that you collect. Label it with your name, date, location it was gathered from, what it was gathered from ( tree, fence, rock..etc. ). Always try to collect physical evidence in the most sterile manner. Use gloves, if no gloves are available use a plastic baggie to cover your hands to avoid cross contamination.

When collecting fluids, like blood, if it still wet use a sterile cotton swab to collect as much as possible, or, if it is dry, dampen it first with sterile water or scrape it with a sterile knife or razor blade. Do not use alcohol as a preservative, this will destroy the DNA. Use a lidded vial to place your samples in, if they are dry use a plain paper envelope. As soon as possible air-dry your sample and place in an envelope. Wet or damp samples will set up bacteria and mold. Using a baggie or vial will contain the moisture and not allow them to dry. To help preserve the sample keep it cool and dry and place in a freezer as soon as possible. Use sterile tweezers to collect tissue samples and treat them in the same manner. Collect hair using sterile tweezers. If there is blood or tissue connected to the hair, treat it in the same manner as you would when collecting those. Hair with no blood or tissue can be placed in baggies or vials as well as paper envelopes.

Use wide clear tape to collect finger, palm or lip prints. Carefully smooth the tape as you lay it across the print to remove any bubbles or wrinkles. Rub the tape to make sure you pick up all of the print and then carefully lift it and place it onto a piece of paper.

Footprints

If you happen to find a potential Bigfoot footprint carefully check the surrounding area for other possible prints.

Measure the print's length and width. If a whole track of prints are found measure the gait between steps by measuring from the back of one heel to the back of the next heel or the same way with toe tip to the next toe tip. Place an item of scale beside the foot print. A photographic scale or other measuring devices would be ideal but if one is not available use something that has a recognizable size such as a pop can or cigarette pack. If more than one print was found place a piece of paper with a number on it by the print also. Photograph the print from every possible angle making sure the whole print, scale, and number is in each photograph. If photographing a whole track of prints try to photograph them from an angle that shows as many prints as possible.

Make notes of the surrounding area that the print was found and include as much information as possible. List the soil type and weather conditions.

If there are broken limbs or trees near the prints, photograph and record all information about them also.

Casting

Plaster is cheap and can be found at any hardware store and can be used in a pinch. We recommend the use of dental stone when making casts. It is stronger and creates a more detailed cast. Dental stone can be purchased at medical supply stores or a quick search on the internet will find a list of suppliers.

If the weather is rainy and wet, cover the tracks with plastic or anything else on hand that will protect them from the elements and keep them from washing out or being damaged until it is safe to make a cast.

Mix your plaster or dental stone with water in a large baggie or bucket. Stir or knead it to make sure it is mixed thoroughly . It should be the consistency of pancake batter or thinner if the track is in dust or very dry soil as not to crush or destroy any details. Over-fill the track if possible so the depth of the track can be easily measured. It would be helpful to have strips of cardboard or other stiff material to form a barrier wall around the print to help contain the over-fill and allow

for a thicker cast. This would be really helpful if it is not a deep print.
Allow 20 – 30 minutes for you cast to setup and dry. This may take longer for larger casts. After you lift your cast, carefully carve your name and date and the corresponding photo number onto the back or topside of it.
Allow the cast to dry an additional 24 hours before carefully attempting to brush away any dirt or debris.

Recording Artifacts
Artifacts in this case are any objects or structures that have suspected to have been made by or used by a Bigfoot creature. Artifacts are usually made from things like sticks or tree limbs, rocks, and/or bones.
If you should happen to find an artifact the first thing to do would be record the location. A GPS unit would be extremely helpful with gathering this information. Drawing a map of the area is also suggested. Carefully measure the artifact's height and width. Photograph the artifact from every angle. You may find it useful to make a log of every photo taken, listing the direction and distance the photo was taken from.
If a nest or bedding area is discovered do not enter the structure until you have determined if the structure is old or is still in use. You can determine the age of the bed by looking to see if the materials used are dry or green, or if there are cobwebs, etc. If it looks like the bed may be fresh or still in use leave the area. You do not want to leave your scent on the bedding or it may cause the creature to abandon it. Keep that area under surveillance to see if the creature returns. If the nest looks old and

abandoned, photograph it and take notes of the area. After you photograph it thoroughly, carefully start taking it apart looking for any physical evidence that may be left behind.

# CHAPTER SEVEN

# Bigfoot Hunting Techniques

By Bigfoot Hunting.com

# General Packing Items

There are many items to consider taking with you on a Bigfoot hunt. Here is a list of some important things to remember:

**General Gear**

- Backpack
- Tent
- Sleeping Bag
- Compass
- Flash Light and Spare Batteries
- Matches of Lighter
- Knife of Multi-tool
- First Aid Kit
- Bug Spray
- Map of the area

**Food, Water, and Cooking**

**Food**

It is important to bring food that is going to last as well as provide a great deal of energy. Some suggestions include:

- Peanut Butter
- Tuna, ham, chicken and beef in cans
- Dried Noodles (Spaghetti, Ramen)
- Dehydrated foods
- Oatmeal
- Crackers
- Hard cheeses
- Candy bars
- Fresh fruit
- Dried fruits and nuts
- Trail mix
- Baby carrots and other veggies
- Powdered milk and Juice Boxes
- Beef jerky and other dried meats

Food items can be left at your basecamp during the day, but remember to bring some snacks with you out on the trail. You'll also want to keep any food out of the reach of wildlife

**Water**

Remember to bring plenty of water on your trip. When packing water, consider the number of people in the group, the ammount of work you will be doing the and temperature. Empty milk or juice jugs work well for carrying water. Bringing smaller containers will work well when you're out on the trail.

## Cooking

You don't need a whole lot of cooking gear but remember to bring plates and utensils for everyone in your group. Disposable dishes work fine. You'll also want a pot to boil water in and even a frying pan will come in handy. You may even want to bring a cooler with ice. The technique of baiting is a common practice when hunting various animals. The idea is to set out a food source that Bigfoot would want to eat, and wait for the creature to come. In order to do this we have to decide exactly what Bigfoot eats. Past examples include the famous skookum body cast being taken when using apples as bait and the jacobs photos were taken when aromatic deer attractant mix and a mineral lick were used. It is thought that the best way to accomplish this is to set out some bait with a trail cam You should then wait at least one week before checking back at the site because it is thought that animals can sense if a human has been in the area recently and may not come within the first couple days.

# Baiting ideas

Here is a list of foods you may want to try using as bait.

- **Plants**

- Barks
- Roots
- Foliage
- Various Grasses
- **Fruits**
  - Apples
  - Plums
  - Grapes
- **Berries**
  - Blackberries
  - Blueberries
- **Vegtables**
  - Corn
  - Onions
- **Nuts**
  - Acorns
  - Walnuts
- **Meats**
  - Deer
  - Squirrel
  - Rabbit
  - Fish

# Pheromone Chips

Pheromone chips are a method used to attract Bigfoot by creating an impregnated scent. The chip is attached to a tree branch, and left for Bigfoot to arrive. They are often accompanied with a trail cam to capture an image. These chips are made up of a mixture of ape and human pheromones. Pheromone chips aren't proven to work for this type of animal, although they are used to attract other wild animals such as deer. Experts say that any primate would be attracted to them, Bigfoot or not.

# Call Blasting

Call blasting, sound blasting, or vocalization is a technique of using recordings of supposed Bigfoot sounds and playing them loudly throughout the forest. This is used as a way to attract a Bigfoot to come near, or to make a sound back. It works much the same way as calling deer, turkey, and other wild animals. The problem with sound blasting is that it's hard to say that the sounds actually came from a Bigfoot, and if they did, would it come near or get scared off.

# Wood Knocking

Wood Knocking is the action of hitting two pieces of wood together to create a noise. The goal in this technique is to get Bigfoot to answer by wood knocking back or by making a sound. This actually does occur with other known apes as a way to communicate over long distances. No Bigfoot has ever actually been seen beating a tree or creating such a sound, but there has been reports of people hearing the sounds of distant wood knocking throughout forests.

# Eating Habits

Bigfoot has been described to have omnivorous eating habits. It is thought that the creature's diet mainly consists of plant materials with some meats. Common foods consist of nuts, fruits, berries, and various animals. Bigfoot is thought to eat any foods that may be available, as well as migrate to find better food sources. Experts say a creature the size of Bigfoot would need a large food supply to sustain a population and require 5000 calories a day

### Food sources
- **Plants**
  - Barks
  - Roots
  - Foliage
  - Various Grasses
- **Fruits**
  - Apples
  - Plums
  - Grapes

- **Berries**
  - Raspberries
  - Blackberries
  - Blueberries
- **Vegtables**
  - Corn
  - Onions
- **Nuts**
  - Acorns
  - Walnuts
- **Meats**
  - Deer
  - Squirrel
  - Rabbit
  - Fish

# Migration Patterns

It has been speculated that Bigfoot may migrate to find better food source One instance of this could be that Bigfoot migrates along with elk and other animals that it uses as a source of food. It was also speculated on the *MonsterQuest* episode, Sasquatch Attack II, after a sighting of Bigfoot near a blueberry harvest. This was after a sighting had occured during a different season the previous year and that Bigfoot may have moved a few hundred miles to find the blueberries.

# Decoys

The technique of using decoys to search for Bigfoot was first seen on History Channel's *MonsterQuest*. This technique is based on the proven fact that it works with many other animals such as deer, turkey, and ducks, so why not Bigfoot? This technique does have its problems though. It's hard to say that a Bigfoot would really want to come near another Bigfoot. It also might realize that the decoy is a fake, based on its intelligence. On the other hand it may be curious to check it out since it is in the area anyway. It may also be seen as a threat.

# CHAPTER EIGHT
## By
### Cryptid Seekers

Cryptid Seekers – Field Equipment: Utilizing Proper Research Tools and Technology

    The brilliant astronomer Dr. Carl Sagan once remarked, "Extraordinary claims require extraordinary evidence." Most of us would readily acknowledge that the existence of creatures like Bigfoot and the Loch Ness Monster in our modern day and age are indeed extraordinary claims. Therefore, a premium must be placed on the importance of definitive, physical evidence in order to prove that cryptids actually do exist. Within the field of zoology, a type specimen is the accepted requisite for classifying a new animal species. This is typically either a live specimen or else sufficient remains that can be examined, measured and compared to known species. Because the collection and preservation of physical evidence presents a wide array of logistical problems, we will address that topic in another section.
    In this section we will be discussing the tools used for the collection of circumstantial evidence, which can help to build a case provided it is of an extremely high standard.

Remote Trail Cameras –

    In recent years, wildlife researchers around the world have benefited from rapid advancements in photographic technology. For example in Indonesia, several new animal species have been discovered over the past decade and in many cases the use of trail cameras have played a major role in these findings. Essentially, these devices are comprised of cameras that have been modified so that the trigger is activated should a certain event occur, specifically a subject moving into the frame of view. The benefits are that these units can be left in remote areas over long periods of time and will only activate at the precise moment when the subject moves into range. Images can be retrieved later, at the convenience of the researcher. Hunters who were attempting to determine the best locations to place their game feeders and hunting blinds employed the first trail cameras. Early models utilized heat sensing motion detectors. This technology has since given way to more accurate passive infared sensors (PIRs), which measure radiating light or energy that is emitted by all animals. This energy is not perceivable by the human eye, but can be detected by certain electronic devices.

Trail cameras are generally constructed to be quite rugged and also to withstand the elements (waterproof), while remaining functional. Typically, these units draw their power from batteries or in some instances from solar panels. Initially trail cameras featured 35mm film as a capture medium, but in recent years most units in use tend to be digital. The quality of the photographic images is measured in megapixels and some models offer better lenses, faster shutters, video capabilities or infared (night vision) for night time or low light (0 lux) conditions. Some models provide a flash for nighttime illumination, but obviously this can alert animals to the presence of the device and startle them. Most utilize an internal clock that will time-stamp the images, in addition to allowing the programmer to determine how many photos will be taken during the course of an event. The outer shells of trail cameras are frequently camouflaged and there is usually an attached strap in order to fasten them to trees.

    An American company called TrailMaster manufactures the best trail monitoring systems. Their products provide superior accuracy and reliability but more importantly, TrailMaster units feature an active infared technology that does not emit any discernable glow, making them less intrusive on the environment. These trail monitors can be used with various types of cameras that ideally are situated in protective casings. TrailMaster products tend to be very expensive, but they are the preferred choice of serious wildlife researchers around the world. A slightly less expensive option is the trail cameras produced by the French company Reconyx, which feature superior images, faster trigger times and an extended battery life. In addition, some Reconyx products feature a patented, discreet sensor that does not emit the red glow typically found in PIR sensors. Other brands that are popular include Cuddleback, (which advertises faster trigger speeds), Stealth Cam and Moultrie.

Personal Cameras –

    Given the rare and elusive nature of cryptids, it is unlikely that most researchers will ever have the opportunity to actually capture a photograph of one.

Nonetheless, every cryptozoologist always carries some type of camera in the field... just in case. Most importantly, cameras are used to document potential evidence including tracks, scat, nests and other anomalous artifacts. Choice of cameras is very much a personal preference, with many variables like brand, quality of lens and resolution, still vs. video, digital vs. film, etc. Obviously it is recommended that a camera be well made and durable, taking into account that it will be exposed to the elements. Also, since your camera goes everywhere, it should be small and lightweight. It goes without saying that a better quality camera with a superior lens will produce better images and that focus is very important for clarity. Most cameras also have a zoom for photographing faraway objects, as well as controls for the f-stop and shutter speed for adjusting to different lighting conditions. There is typically a flash mechanism as well. Many of the Sony brand cameras have a feature called Night Shot, which is a rudimentary form of night vision that can be used for less intrusive filming at night, though the resulting images are black and white. The advantage of shooting video as opposed to stills is that it provides more information, including audio and a timeline. Make sure you are intimately acquainted with how your camera works, since any encounter would more than likely happen very fast and your adrenalin would be pumping.

Digital Audio Recorders –

Frequently, cryptozoological field research involves listening for vocalizations or other sounds that are associated with a particular cryptid. Since most animals can project quite well in the wild, you have a good chance of hearing something intriguing if you are within range. Recordings of audio that cannot be easily identified can make for compelling evidence. There are a number of battery operated, digital, handheld audio recorders on the market. Most of them have onboard condenser type microphones that are superior to human hearing and capable of reproducing a wide range of frequencies. Costlier units will feature better microphones with a greater range (sensitivity) and more clarity. Typically, these recorders are capable of 24-bit/96 KHz sample resolution and use .wav format, which is easily transferable to a PC. Various kinds of audio software allow for enhancements and analyzation. It is always preferable to record at the highest resolution possible for optimum quality and also to maximize your input levels for a stronger signal. But, be careful not to clip or distort your input by having it up too high.

GPS –

Global Positioning Satellite has greatly impacted the world that we live in. It tells us with great precision exactly where we are at any given moment. Battery-operated, handheld GPS units that are specifically made for sportsmen are available. These models are rugged and waterproof and typically contain database maps of wilderness areas and waterways. As a researcher, there are many benefits to knowing your exact Earth coordinates at any given moment. Virtual markers can be used in order to designate important locations, such as those where a cryptid has been sighted or where evidence has been found. Also, these devices make navigating in the woods much easier and safer. But, GPS may not always be effective in the thick woods or in valleys where trees and mountains can obstruct satellite signals.

Binoculars –

Field glasses, as they are sometimes called, are standard issue for any nature researcher. Essentially two telescopes affixed together, they allow the viewer to see objects that are a good distance way in three dimensions. Like cameras, you get what you pay for. Some field glasses can cost as much as a car, because of the quality of the lenses/glass and engineering. Imagine losing them in the woods though! A decent pair of binoculars can be purchased for around $30. Binoculars are measured in terms of their ability to magnify. For example with 10x25, the first number represents the level of magnification, so that 10 would mean that an object looks ten times closer than it actually is. The second number measures the diameter of the outer (objective) lens, so that 25 would mean twenty-five millimeters across. With higher levels of magnification, the image does not look as stable to the human eye, though better quality units generally address this issue. Most binoculars provide focus and depth controls. Some use antireflection coatings to cut down on glare.

Flashlights –

Like many animals, most cryptids are believed to be primarily nocturnal. Subsequently, a great deal of research is done at night when creatures are most active. It is therefore important to keep a flashlight handy at all times. Advancements in technology have made for longer lasting batteries and bulbs with better illumination so good flashlights are easy to come by these days. Many cryptozoologists wear headlamps, which are essentially flashlights attached to a strap that are worn on the head. These enable your hands to be free for other tasks. Most headlamps have standard, xenon and discreet, red or green low light settings. There are also battery-charged floodlights that can literally light up the forest at night if needed. These units are rated by something called candlepower, which can be as high as three million in strength. Spotlights can be good defensive tools too, since they can stun or scare off threatening animals. Never look into one of these devices at close range or you could seriously damage your eyes.

Night Vision/Thermal Imaging –

Originally developed by the military, these technologies have become extremely sought after by cryptozoological researchers, though units can be quite expensive and sometimes hazardous. Basically, night vision utilizes infared light sources that are unperceivable to the human eye, in order to increase visibility in the dark. There have been four generations of NV technology over the years, with Gen 1 being very inexpensive and rudimentary, Gen 2 slightly better, Gen 3 fairly impressive but expensive and Gen 4 generally reserved for military use only. The resulting images are two-tone, with a greenish hue along with various shades of gray or black. One should be careful not to purchase from the plethora of relatively cheap Eastern European products on the market. They cannot be calibrated properly and can seriously damage your eyesight. Night vision viewers that can be found in children's science kits are pretty cool for the money and much safer to use.

Thermal Imaging or FLIR, actually measures the temperatures of solid objects and air masses, allowing the user to differentiate by projecting images on a viewer or camera. These devices are frequently used in both industrial and rescue situations and are quite remarkable for detecting life forms in dark areas, though they typically cost thousands of dollars. The resulting images are not always that finite, due to the nature of these units.

Listening Devices –

Parabolic microphones are long range listening devices, which amplify distant sounds in order to make them more audible. These units can be quite useful when listening for noises that are a good distance away. By design, these bowl shaped mics are usually bulky and most produce generally low quality audio because of how sounds waves travel. But, they can be useful tools when trying to locate your quarry. There are also so-called bionic ears, which are basically ear buds attached to a small, parabolic microphone. They are inexpensive, but not really that effective. One should always be careful when placing headphones or other amplified devices close to the eardrum, as your hearing could be damaged if you don't watch your volume.

Communications –

When working with other researchers, one objective is to cover as much ground as possible. Therefore it is important for team members to have a strong and reliable line of communication when separated by distance. Cell phones aren't always an option in remote areas where service isn't available. Walkie-talkies or shortwave, handheld radios have been in existence since World War II and haven't changed too terribly much since that time. Be sure to purchase units with a long range, (preferably 12 miles) and always make sure you and your team members have agreed upon a designated channel for all correspondence. Remember to always act responsibly when broadcasting in a public forum
Game Calls –

Used by hunters for centuries, game calls are basically devices that broadcast the sounds of known animals, in order to elicit a response from potential prey in the vicinity. These can be as simplistic as a fashioned, hollow piece of wood or plastic, or as advanced as a digital playback system with a database of audio files and a powered speaker system. The latter units have been used in Bigfoot research by many investigators who typically broadcast recorded vocalizations that have presumably been made by these creatures. Fox Pro manufactures some high quality models. Game calls can be fun to experiment with, since certain distress calls will draw in a wide range or wildlife. Using customized, non-threatening recordings (for example the sound of children laughing) might peak the curiosity of sentient creatures.

Sonar –

Looking for evidence of hidden life forms under the water presents obvious logistical obstacles. The best technology available at this stage is sonar, which essentially uses a technology whereby sound waves are projected through the depths. The resulting reflections paint a picture of what lies below that can be viewed and also recorded. Fishermen routinely use rudimentary forms of sonar in order to locate fish and also to examine the topography under their boats. Side scan sonar utilizes a cylindrical device that is towed alongside the boat. It can cover a wider area with impressive results, but it's also more expensive.

To Contact a Professional Sketch Artist for your sighting please contact Pete Travers @
Email: pete@thepaintedcave.com

Pete Travers is working with close-range eyewitnesses to develop detailed sketches of bigfoots / sasquatches. He has developed several composite sketches so far, and is seeking contact with other close-range eyewitnesses to develop more sketches.

The sketches he has developed so far are not displayed on the web, in order to help preserve the clarity of individual witness memories. The actual development of the sketches occurs online, however. Peter creates initial sketches, then e-mails the sketch to the witness, then refines the sketch based upon the input of the witness. The process is repeated until the sketch looks as close as possible to what the witness observed. It is not a continuous process. It is separated into stages to make it as convenient as possible. Nor is it very time consuming. It is, however, very helpful for accumulating knowledge about these animals.

If you are close-range eyewitness and you are willing to work with a sketch artist, please contact Pete Travers at pete.g.travers@gmail.com and describe your encounter.

About Travers Work
The step by step process

## Sasquatch Head Sketch Session

Location: Colorado
Report: 2110

**Initial client comments**
**Head shape:** Roundish. The neck was thick and not that long.
**Hair:** All over. Fur all the way up to the lips and eyes.
**Skin:** Light brown.
**Eyes:** Eyes were large and black. Rounder than a normal humans. Less fur around the eyes and it was lighter.
**Nose:** Just like the Patty sketch.
**Ears:** Don't remember the ears.
**Mouth:** Fur all of the way up to the lips.
**Other:** Looked alot like the Patty sketch. No area without fur.

## *Iteration One*

**Witness comments to iteration one:** Hair longer. Visible mouth a little less wide -- maybe 2/3 of the dark line on the left side slightly less visible. Neck slightly longer (Maybe 1/8" lower as it enlarges on my screen -- or about 1/2 way through the dark locks on the right). It seems that the shape of the brow was slightly broader at the eyes and the eyes just a little farther apart but leave that until the other changes. As I hold up fingers and thumbs to see the neck and mouth differently those seem the most crucial. The eyes are really good. As I mentioned when we talked, the fur around the eyes and on the nose of the individual I saw was sparcer and shorter but there and I think you caught that. If there is any change on the fur, don't darken those areas. Also, I know this sounds stupid, but the expression of the eyes was both very intense (alert, inquisitive) and, forgive the subjectivity, very compassionate.

## *Iteration Two*

**Witness comments to iteration two:** The face is there -- I hope you can use this sketch with a modification because you've captured something about it that I'm afraid of loosing. The most important change is in the shape of the head which is broader (rounded) at the eyebrow, eye and cheek level. The upper head is good. Over the right eye on your sketch is a line that comes from the bridge of the nose over above the eyebrow. From the height of the top of that line is where the head needs to begin to be broader and the outline would come back to your drawing right between the height of the nose and mouth and then the shoulders out from there. (It is a rounding of that area) The amount of broadening of the side of the head is about the depth of the upper lip or slightly less. The shoulders start out at a height just above the mouth (I know, I know...last time I told you the neck was longer...) If you want I can fax you what I did to the sketch that made it click. The face is very like what I saw. There is something about the nose that I want to change but I'm incapable of seeing what it is and I feel comfortable that this looks like what I saw.

## *Iteration Three*

**Witness comments to iteration three:** Bring the rounded area on the right side down to the level of the mouth and it's there. This looks like what I saw.

## *Iteration Four (Final)*

**Witness comments to iteration four:** I'm comfortable with this sketch..

## *Finished Sketch*

## *Chapter Nine*
## *Sasquatch Forum Sites:*

- Alabama Bigfoot Research Forum
- Bigfoot Discussions
- Bigfoot Forums
- Bigfoot, The North American Ape Forum
- Cryptovideography Forum
- GCBRO Message Board
- JREF - James Randi Educational Foundation Forum
- Kiamichi Giants Bigfoot Research Message Board
- NESRA - Northeast Sasquatch Researchers Association Forum
- North American Bigfoot Forum (Stan Courtney)
- Pennsylvania Bigfoot Society Forum (PBS)
- Sasquatch Watch Forum (Billy Willard / Tom L / DB Donlon)
- Squatchdetective Forum (Steve Kulls)
- The Search for Bigfoot Forum
- Timberline Bigfoot Group Forum (TBG)

## *Bigfoot Information Sites & Organizations:*

**Adirondack Bigfoot Club (NY)**

**Alabama Bigfoot Research**

**Alliance of Independent Bigfoot Researchers (AIBR)**

**American Bigfoot Society (ABS)**

**Bigfoot Au Quebec (in French)**

**Bigfoot Discovery Project (CA)**

**Bigfoot Encounters (Bobbie Short)**

**Bigfoot Exterminators (Don Barone & Denver Riggleman - ESPN)**

**Bigfoot: Fact or Fantasy? (Roger Thomas)**

**Bigfoot Sketch Project (Pete Travers)**

**Bigfoot Times (Newsletter Site - Daniel Perez)**

**Canadian Sasquatch Research Organization (formerly known as the Western Canadian Sasquatch Research - WCSRO)**

**Central Ohio Bigfoot Research (C.O.B.R.)**

**Cryptomundo.com**

**Eastern Ohio Bigfoot Research Center**

**Georgia Bigfoot**

**Indiana Bigfoot Awareness Group**

**International Bigfoot Society**

Maryland Bigfoot

Michigan Bigfoot

Mid-America Bigfoot Research Center (MABRC)

Monsters & Myths - Cryptid Blog

New Jersey Bigfoot Reporting Center

North America Bigfoot Search (David Paulides - CA)

Northern Sasquatch Research Society (Bill Brann - NY)

NESRA - North East Sasquatch Researchers Association (Mid-Atlantic & Northeast USA & Northeast Canada)

NESRA Member Blogs

Oregon Bigfoot (Autumn Williams)

Pennsylvania Bigfoot Society (PBS)

Sasquatchers.com (Alabama Research Site)

Sasquatch Information Society (Seattle)

Sasquatch Watch VA (Billy Willard / Tom L / DB Donlon)

Search For Bigfoot (Melissa Hovey - Texas)

Shenandoah Valley Sasquatch Foundation

Skunk Ape Research Foundation Of Western New York

Sierra Sasquatch Research Group (Tahoe Basin CA)

Southern Bigfooter Blog - Bigfoot On Ice

Southern New England Research Organization

**(SNEBRO)**

**South Alabama Bigfoot Research**

**SquatchDetective.com (Steve Kulls - NY)**

**Squatchtopedia - The Bigfoot Wiki**

**SSA - Southeast Sasquatch Association (Henry May)**

**SSA Blog**

**StanCourtney.com (Illinois Research Site)**

**Stocking Homonid Research (Diane & Donna Stocking - FL and OR site)**

*Bigfoot / Sasquatch Radio Shows:*

- **Bigfoot Field Guide Radio Show**
- **Bigfoot Quest - Radio Show**
- **The Gray Matter - Radio Show**
- **Let's Talk Bigfoot - Radio Show**
- **The Sasquatch Experience & The Sasquatch Triangle - Radio Shows**
- **Sasquatch Watch Radio**

To contact author: cryptidseekers@hotmail.com

# Chapter Ten
# BIGFOOT NAMES

**Abominable Snowman** (Nepal and Tibet)
**Abonesi** (Africa)
**Agogwe** (Africa)
**A hoo la huk** or **A hoo la hul** (America- Yup'ik Indians in Alaska)
**Alma** (Russia)
**Amajungi** (Africa)
**Amomongo** (Philipines)
**Atahsaia** (America- Zuni Indians in New Mexico)
**At'at'ahila** (America- Chinookan Indians in numerous Pacific Northwest states)
**Atshen** (Canada- Tete-de-Boule or Atikamek Indians)
**Atoe Pandak** (Sumatra)
**Atoe Rimbo** (Sumatra)
**Atshen** (Canada- Tete-de-Boule or Atikamek Indians)
**Ban-manush** (Bangladesh)
**Bardin Booger** (America- Florida)
**Barmanou** (Afghanistan and Pakistan)
**Batutut** (Sumatra)
**B'gwas** (Canada- Haisla Indians)
**Bigfoot** (America- used interchangeably with Sasquatch)
**Big Muddy Monster** (America- Illinois)
**Booger** (America- Florida)
**Boqs** (America and Canada- Bella Coola Indians)
**Bukwas** (Canada- Kwakwaka'wakw Indians)
**Chenoo** (Canada- Micmac Indians)
**Chickly Cudly** (America- Cherokee Indians in North Carolina)

**Chiha tanka** or **Chiye tanka** (America- Lakota and Sioux Indians from the Dakotas)
**Chimanimani** (Africa)

**Chittyville Monster** (America- Illinois)
**Choanito** (America- Wenatchee Indians in Washington)
**Chuchunaa** (Russia)
**Devil Monkey** (America- midwest)
**Djenu** (Canada- Micmac Indians)
**Doko** (Africa)
**Dzu-Teh** (Nepal and Vietnam)
**Ebu Gogo** (Indonesia)
**El-Ish-kas** (America- Makah Indians in Washington)
**Esti Capcaki** (America- Seminole Indians in numerous states)
**Farmer City Monster** (America- Illinois)
**Fear Liath** (Scotland)
**Forest Man** (Vietnam)
**Fouke Monster** (America- Arkansas)
**Ge no'sgwa** (America and Canada- Seneca and Iroquois Indians)
**Get'qun** (America- Lake Lliamna Indians in Alaska)
**Gilyuk** (America- Nelchina Plateau Indians in Alaska)
**Gin-sung** (China)
**Goegoeh** (Sumatra)
**Gogit** (America- Haida Indians in Alaska)
**Goo tee khi** or **Goo tee khl** (America- Chilkat Indians in Alaska)
**Gougou** (Canada- Micmac Indians)
**Grassman** (America- Ohio)
**Greyman** (Scotland)
**Gugu** (Sumatra)
**Gugwe** (Canada- Micmac Indians)

**Hairy Man** (America- Eskimo Indians in Alaska)
**Hecaitomixw** (America- Quinault Indians in Washington)
**Hibagon** (Japan)
**Ijiméré** (Africa)
**Iktomi** (America- Plains Indians in numerous states)
**Ijaoe** (Sumatra)
**Isnashi** (South America)
**Kakundakári** (Africa)
**Kala'litabiqw** (America- Skagit Indians in Washington)
**Kapre** (Philipines)
**Kashehotapalo** (America- Choctaw Indians in numerous southeastern states)
**Kecleh-kudleh** (America- Cherokee Indians in numerous southeastern states)
**Kiwakwe** (America- Penobscot Indians in Maine)
**Koakwe** (Canada- Micmac Indians)
**Kokotshe** (Canada- Tete-de-Boule or Atikamek Indians)
**Kushtaka** (America- Tlingit Indians in Alaska)
**Lake Worth Monster** (America- Texas)
**Lariyin** (America- Hare Indians in Alaska)
**Loo poo oi'yes** (America- Miwuk Indians in California)
**Madukarahat** (America- Karok Indians in Califonia)
**Manabai'wok** (America- Menominee Indians in Wisconsin)
**Mande Barung** (India)
**Mapinguari** (Brazil)
**Marimonda** (South America)
**Matlose** (Canada- Nootka Indians)
**Mau** (Africa)
**Mberikimo** (Africa)

**Meh-Teh** (Nepal and Vietnam)
**Memegwicio** (America- Timagami Ojibwa in Minnesota)
**Mesingw** (America- Lenni Lenape Indians in Delaware and New Jersey)
**Metch-kangmi** (Nepal and Tibet)
**Miitiipi** (America- Kawaiisu Indians in California)
**Misinghalikun** (America- Lenni Lenape Indians in Delaware and New Jersey)
**Mogollon Monster** (America- Arizona)
**MoMo** (America- southern region/Missouri)
**Mono Grande** (South America)
**Myakka Ape** (America- Florida)
**Na'in** (America- Gwich'in Indians in Alaska)
**Nalusa Falaya** (America- Choctaw Indians in Mississippi)
**Nantiinaq** (America- Kenai Peninsula Indians in Alaska)
**Nant'ina** (America- Dena'ina Indians in Alaska)
**Neginla eh** (America and Canada- Alutiiq and Yukon Indians)
**Ngagi** (Africa)
**Nguoi Rung** (Vietnam)
**Niaka-Ambuguza** (Africa)
**Nuk-Luk** (America- northwest region)
**Nu'numic** (America- Owens Valley Paiute Indians in California)
**Nun Yunu Wi** (America- Cherokee Indians from numerous southeastern states)
**Nyalmo** (Himalayan region)
**Oh Mah** (America- Hoopa Indian)

**Omah** (America- Yurok Indian)

**Old Yellow Top** (Canada)
**Orang Gugu** (Sumatra)
**Orang Letjo** (Sumatra)
**Orang Mawas** (Malaysia)
**Orang Pendek** (Indonesia and Sumatra)
**Ot ne yar heh** (America- Iroquois Indians in numerous states)
**Qah lin me** (America- Yakima and Klickitat Indians in Washington)
**Qui yihahs** (America- Yakima and Klickitat Indians in Washington)
**Rougarou** or **Rugaru** (America and Canada- Ojibway Indians)
**Saskahavas** or **Sasahevas** (Canada- Halkomelem Indians)
**Saskets** (America- Salishan/Sahaptin Indians in Oregon and Washington)
**Sasquatch** (Canada- used interchangeably with Bigfoot)
**Sc'wen'ey'ti** (America- Spokane Indians in Washington)
**Seatco** (America- Yakima, Klickitat, and Puyallup Indians in Washington)
**Seat ka** (America- Yakima Indians in Washington)
**Sedabo** (Sumatra)
**Sedapa** (Sumatra)
**See'atco** (America and Canada- Salish Indians)
**Seeahtkch** or **Seeahtik** (America- Clallam Indians in Washington)
**Selahtik** (America- Clallam Indians in Washington)
**Séhité** (Africa)
**Siatcoe** (America- Clallam Indians in Washington)
**Skanicum** (America- Colville Indians in Washington)
**Skookum** (America and Canada- Chinook Indians)

**Skukum** (America- Quinault Indians in Washington)
**Skunk Ape** (America- southern region/Florida)
**Slalakums** (Canada- Upper Stalo Indians)
**Sne nah** (America and Canada- Okanogan/Okanogon Indians)
**So'yoko** (America- Hopi Indians in Arizona)
**Steta'l** (America- Puyallup and Nisqually Indians in Washington)
**Ste ye mah** (America- Yakima Indians in Washington)
**Stink Ape** (America- Florida)
**Strendu** (Canada- Huron and Wyandot Indians)
**Swamp Ape** (America- southern region/Florida)
**Tah tah kle' ah** (America- Yakima Indians in WA and Shasta Indians from CA)
**Tano Giant** (Ghana)
**Teh-Ima** (India)
**Tenatco** (Canada- Kaska Indians)
**Tjutjuna** (Russia)
**Toké-Mussi** (America- Oregon)
**Tokoleshe** (Africa)
**Tornit** (America and Canada- Inuit Indians)
**Toylona** (America- Taos Indians in New Mexico)
**Tsadjatko** (America- Quinault Indians in Washington)
**Tse'nahaha** (America- Mono Lake Paiute Indians in California)
**Tsiatko** (America- Puyallup and Nisqually Indians in Washington)
**Tso apittse** (America- Shoshone Indians in numerous states)
**Tsonaqua** (Canada- Kwakwaka'wakw Indians)
**Ucumar** (Argentina)
**Uhang Pandak** (Sumatra)

**Ujit** (Vietnam and Borneo)
**Umang** (Sumatra)

**Umang** (Sumatra)
**Urayuli** (America- Eskimo Indians in Alaska)
**Weendigo** (America and Canada- Algonquian speaking Native American Indians)
**Weetigo** (America and Canada- Algonquian speaking Native American Indians)
**Wendigo** (America and Canada- Algonquian speaking Native American Indians)
**Wetiko** (America and Canada- Cree Indians)
**Windago** (America and Canada- Algonquian speaking Native American Indians)
**Winstead Wild man** (America- Connecticut)
**Witiko** ((Canada- Tete-de-Boule or Atikamek Indians)
**Wittiko** (America and Canada- Algonquian speaking Native American Indians)
**Woods Devil** (America- New Hampshire)
**Wookie** (America- Louisiana)
**Wsinkhoalican** (America- Lenni Lenape Indians in numerous states)
**Xi'lgo** (America- Nehalem and Tillamook Indians in Oregon)
**Yahoo** (America- North Carolina)
**Yahyahaas** (America- Modicum Indians in Oregon and California)
**Yayaya-ash** (America- Klamath Indians in Washington)
**Yé'iitsoh** (America- Navajo Indians in numerous states)
**Yeren** (China)
**Yeti** (Nepal and Tibet)
**Yi'dyi'tay** (America- Nehalem and Tillamook Indians in Oregon)

**Yohemiti** (America- Miwok Indians)
**Yowie** (Australia)

Made in the USA
Middletown, DE
17 April 2018